David Lochbaum is the director of the Union of Concerned Scientists' Nuclear Safety Project and the author of *Nuclear Waste Disposal Crisis.*

Edwin Lyman is a senior scientist in the Global Security Program of the Union of Concerned Scientists.

Journalist **Susan Q. Stranahan** was a member of the *Philadelphia Inquirer* team awarded a 1980 Pulitzer Prize for its coverage of the Three Mile Island accident.

The **Union of Concerned Scientists** is the leading science-based nonprofit working for a healthy environment and a safer world.

FUKUSHIMA

The Story of a Nuclear Disaster

David Lochbaum, Edwin Lyman, Susan Q. Stranahan,
and the Union of Concerned Scientists

THE NEW PRESS

NEW YORK
LONDON

Requests for permission to reproduce selections from this book should be mailed to:
Permissions Department, The New Press, 120 Wall Street, 31st floor, New York, NY 10005.

First published in the United States by The New Press, New York, 2014
This paperback edition published by The New Press, 2015
Distributed by Perseus Distribution

ISBN 978-1-62097-084-3 (paperback)
ISBN 978-1-62097-118-5 (e-book)

LIBRARY OF CONGRESS CATALOGING-IN-PUBLICATION DATA
Lochbaum, David A.
    Fukushima : the story of a nuclear disaster / David Lochbaum, Edwin Lyman,
Susan Q. Stranahan, and the Union of Concerned Scientists.
        p.   cm.
    Includes bibliographical references and index.
    ISBN 978-1-59558-908-8 (hc. : alk. paper)
    1. Fukushima Nuclear Disaster, Japan, 2011.   2. Nuclear power plants—Accidents—Japan—
Fukushima-ken.   I. Union of Concerned Scientists.   II. Title.
    TK1365.J3L63 2014
    363.17'990952117—dc23   2013035284

The New Press publishes books that promote and enrich public discussion and understanding of
the issues vital to our democracy and to a more equitable world. These books are made possible
by the enthusiasm of our readers; the support of a committed group of donors, large and small;
the collaboration of our many partners in the independent media and the not-for-profit sector;
booksellers, who often hand-sell New Press books; librarians; and above all by our authors.

www.thenewpress.com

Composition by dix!
This book was set in Minion

Printed in the United States of America

10   9   8   7   6   5   4   3   2   1

CONTENTS

Introduction to the Paperback Edition vii

1. March 11, 2011: "A Situation That We Had
   Never Imagined" 1

2. March 12, 2011: "This May Get Really Ugly ..." 34

3. March 12 Through 14, 2011: "What the Hell Is Going On?" 55

4. March 15 Through 18, 2011: "It's Going to Get Worse ..." 79

5. Interlude—Searching for Answers: "People ... Are
   Reaching the Limit of Anxiety and Anger" 103

6. March 19 Through 20, 2011: "Give Me the Worst Case" 121

7. Another March, Another Nation, Another Meltdown 141

8. March 21 Through December 2011: "The Safety
   Measures ... Are Inadequate" 155

9. Unreasonable Assurances 182

10. "This Is a Closed Meeting. Right?" 204

11. 2012: "The Government Owes the Public a Clear and
    Convincing Answer" 222

12. A Rapidly Closing Window of Opportunity 244

Appendix: The Fukushima Postmortem:
What Happened? 263

Glossary 269

Key Individuals   277

U.S. Boiling Water Reactors with "Mark I" and "Mark II"
Containments   279

Notes and References   281

Index   293

INTRODUCTION TO THE PAPERBACK EDITION

Nearly four years have passed since the accident at the Fukushima Daiichi nuclear plant in Japan. Since that time, the long-term environmental and economic consequences of the accident and the magnitude of the effort it will take to address them have come into sharper focus. While Tokyo Electric Power Company (TEPCO) has almost completed removal of the spent fuel assemblies from their precarious perch in Unit 4, the inability to contain the vast and growing quantity of contaminated water at the plant site has emerged as one of the biggest problems that Japan is facing. Water being pumped into the reactors to cool the damaged cores continues to leak out through cracks in the containment structures while hundreds of gallons of groundwater flow under the site every day, washing radioactive contamination into the sea. TEPCO does not have the capacity to collect and treat all of this water. It is constructing a system to freeze a mile-long wall of soil around the contaminated area in order to divert groundwater flow, but that may not work. Preliminary attempts to freeze a much smaller area have failed.

Although more details about the accident and its aftermath have come to light, the fundamental elements of this disaster remain the same: a trio of reactor cores in meltdown; an extended loss of all power; vulnerable pools of deadly spent fuel at risk of boiling dry; radiation threatening large swaths of Japan, including Tokyo, the world's most populous metropolis, and potentially even parts of the United States. Images of reactor buildings exploding, stories of heroic efforts to save the plant, poignant accounts of families uprooted from their homes and heritage, and communities rendered uninhabitable will remain vivid in the public consciousness for years to come.

The story of Fukushima Daiichi is a larger tale, however. It is the saga of a technology promoted through the careful nurturing of a myth: the myth of safety. Nuclear power is an energy choice that gambles with disaster.

Fukushima Daiichi unmasked the weaknesses of nuclear power plant design and the long-standing flaws in operations and regulatory oversight. Although Japan must share the blame, this was not a Japanese nuclear accident; it was a nuclear accident that just happened to have occurred in Japan. The problems that led to the disaster at Fukushima Daiichi exist wherever reactors operate.

The staff of the U.S. Nuclear Regulatory Commission (NRC) appear to have finally accepted this as well, concluding in a November 2013 report that even if Japan had the same regulatory framework as the United States prior to the accident, there is no assurance "that the Fukushima accident and associated consequences could or would have been completely avoided."

Although the accident involved a failure of technology, even more worrisome was the role of the worldwide nuclear establishment: the close-knit culture that has championed nuclear energy—politically, economically, socially—while refusing to acknowledge and reduce the risks that accompany its operation. Time and again, warning signs were ignored and near misses with calamity written off.

Important lessons from Fukushima Daiichi continue to emerge. TEPCO now believes that core damage at Unit 3 was more extensive than it had previously estimated because the emergency core cooling system in the Unit 3 reactor stopped working earlier than it had thought. Consequently, the core became uncovered around 2:30 a.m. on March 13, rather than 9:00 a.m., as we reported in the book. And TEPCO is now convinced that a containment failure at Unit 2 on March 15, coupled with an unfavorable weather pattern, was the cause of the extensive radiation contamination to the northwest of the site. However, much still remains unknown about what happened inside the Japanese nuclear plant.

Also unclear is the magnitude of the long-term effects of radiation releases on human health and the environment, as well as the ultimate economic impact. Based on a 2014 estimate of the total radiation dose to the Japanese public for eighty years after the accident by the UN Scientific Committee on the Effects of Atomic Radiation, a few thousand cancer deaths could be expected. However, the ultimate human toll depends on the fate of the more than one hundred thousand evacuees who remain displaced because their homes are contaminated. The Japanese authorities could limit future radiation exposures by enforcing strict cleanup standards before allowing the evacuees to return to their homes, but instead they are declaring areas safe where radiation levels are ten times greater than normal background levels.

One thing is certain: absent the valiant and tireless efforts of many at Fukushima Daiichi, the consequences could have been much, much worse.

Fukushima Daiichi provided the world with a sobering look at a nuclear accident playing out in real time. Like previous accidents at Three Mile Island and Chernobyl, the events that began on March 11, 2011, defied the computer simulations and glib assurances of nuclear power promoters that this form of energy is a prudent and low-risk investment. It cannot hope to fulfill that claim absent an unwavering and uncompromising commitment to safety.

Excuses about this accident flew almost from the outset. Nobody had predicted an earthquake this large. Nobody had expected a huge tsunami to flood a low-lying coastal nuclear plant. Nobody had envisioned an accident involving multiple reactors. Nobody had assumed such an event could involve a loss of power for more than a few hours. But all of this did happen, and within a matter of hours the assurances of nuclear safety were revealed as a fallacy.

Even so, many in the United States, Japan, and elsewhere are pushing hard to defend the status quo and hold fast to the assertion that severe accidents are so unlikely that they require scant advance planning. Those who might differ with that view—among them the tens of thousands of Japanese whose lives have been radically altered—have thus far largely been shut out of the public policy debate over nuclear power.

In Japan, the conservative government under Prime Minister Shinzo Abe continues its push to restart Japan's nuclear plants as rapidly as possible. The Nuclear Regulation Authority (NRA), which was created in the wake of the accident to bolster the credibility of the Japanese nuclear regulatory system, has begun the process with its approval of the restart of the Sendai plant on the southern island of Kyushu, despite its location thirty miles away from an active volcano. Questions linger about the independence of the NRA, as Abe replaces its initial members with pronuclear individuals with close industry ties. The government simply seems in denial about the very real potential for another catastrophic accident.

In the United States, the NRC has also continued operating in denial mode. It turned down a petition requesting that it expand emergency evacuation planning to twenty-five miles from nuclear reactors despite the evidence at Fukushima that dangerous levels of radiation can extend at least that far if a meltdown occurs. It decided to do nothing about the risk of fire at overstuffed spent fuel pools. And it rejected the main recommendation of its own Near-Term Task Force to revise its regulatory framework. The NRC and the industry instead are relying on the flawed FLEX program as a panacea for any and all safety vulnerabilities that go beyond the "design basis." This should provide little comfort to the public living near the dozens of nuclear plants around the United States that are susceptible to earthquakes and floods far larger than they were originally designed to withstand.

Other methods of generating energy also carry risks in terms of environmental costs as well as human health and safety impacts. But that is no excuse for continuing to hold nuclear power only to the inadequate safety standards that made the Fukushima disaster possible. Nuclear energy is an unforgiving technology, and the consequences of a mistake can be catastrophic.

This book is a collaboration. It weaves a detailed explanation of what went wrong inside the crippled nuclear plant together with a narrative of events taking place in the halls of government in Tokyo and the emergency operations center of the NRC. It reveals how those responsible for protecting the public in the United States and Japan were caught unprepared and often helpless.

Of equal importance, the book describes the nuclear establishment's multidecade effort to weaken safety rules and regulatory oversight, especially in the United States. In doing so, this book answers the question heard so often in this country in the aftermath of March 11, 2011: can it happen here? The answer is an unequivocal yes.

Fukushima: The Story of a Nuclear Disaster also addresses another critical question: how can we work toward ensuring that it never does?

The minute-by-minute account of events at Fukushima Daiichi, as revealed by conversations among the nuclear experts, gives the average reader a rare glimpse into just how complex a nuclear accident can be—and how ill-equipped regulators and the industry are to deal with one, no matter where it takes place. Absent significant upgrades in nuclear operation and regulation, it will be only a matter of time until the world watches as another Fukushima unfolds.

Profound thanks go to Mary Lowe Kennedy, whose skills as an editor run very deep. She jumped into a book about nuclear safety with enthusiasm, with a dedication that never flagged, and with the shared goal of making this a story understandable to all who remember Fukushima and want to learn more.

Special thanks also go to three members of the Union of Concerned Scientists' staff: Stephen Young for taking the initiative to launch the project and Lisbeth Gronlund and David Wright for their counsel and commentary. Peter Bradford generously provided his insights, as did Charles Casto, whose firsthand knowledge of the accident added a valuable dimension to the book. Yuri Kageyama of the Tokyo bureau of the Associated Press generously volunteered her time, helping with translation and research.

Thanks also to Katja Hering for her support and assistance with obtaining documents, and to Rhoda and Andrea Lyman for their support for this project.

Edwin Lyman and Susan Q. Stranahan
October 2014

1

MARCH 11, 2011:
"A SITUATION THAT WE HAD NEVER IMAGINED"

At 2:49 p.m., the chandeliers began to sway in an ornate hearing room of the Diet Building, home to Japan's parliament. Prime Minister Naoto Kan glanced nervously toward the high ceiling. Legislators darted about, their voices rising as the shaking increased. One speaker advised everyone to duck under the desks. Aides hurried to the prime minister's side, uncertain where safety actually lay.

NHK, Japan's public broadcasting system, was televising the Kan hearing. As an unsteady camera captured the confusion in the hearing room, the network was receiving an earthquake alert from the Japan Meteorological Agency. Ground-motion sensors along the coast near Sendai, north of Tokyo, had picked up seismic activity offshore. Based on those readings, the agency estimated that a magnitude 7.9 quake was likely. In thirty seconds it dispatched a warning to residents along the northeastern coast. Personal cell phones lit up. Businesses, schools, hospitals, and the news media all received alerts; twenty-four bullet trains operating in the region glided to a halt.

Within ninety seconds, NHK had interrupted its coverage to provide early details about the quake. Almost instantly, TV screens across Japan began displaying emergency information.

In Sendai, an NHK cameraman hopped aboard the helicopter kept on permanent standby for the network and lifted off from the airport. It would be perhaps the last aircraft to depart the facility before the earthquake crisis took a dramatic turn for the worse. The footage shot from that helicopter would soon be replayed again and again around the world.

Japan sits atop one of the most earthquake-prone regions of the world, shaken by more than a thousand tremors each year. The ancient Japanese believed a giant catfish lay buried beneath the islands. When the creature thrashed about, the earth moved. Modern scientists have their own explanation: several massive plates of the earth's crust abut each other around the islands of Japan, shifting continuously a few inches per year. If their movement is impeded, stress builds until it is released in a seismic event. That event may be barely detectable—or it can be catastrophic.

Along the northeastern coast, the westward-moving Pacific Plate is forced beneath the North American Plate that holds most of northern Japan. This downward movement along a seam in the earth's surface is known as subduction. It is responsible for producing some of the world's largest earthquakes.

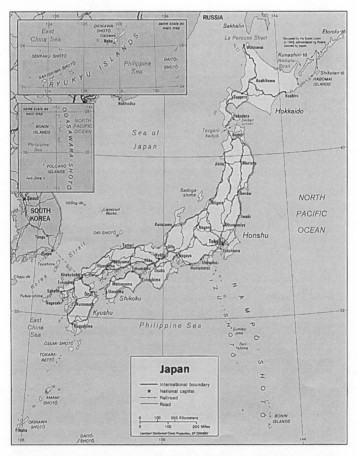

A map of Japan showing the major cities. *Perry-Castañeda Library Map Collection, University of Texas*

Although northeastern Japan is no stranger to earthquakes, seismologists long believed that a subduction quake was unlikely to happen there because the 140-million-year-old Pacific Plate was sliding downward smoothly, never creating a huge buildup of stress. Unbeknownst to them, however, sections of the two plates had locked together, possibly for as long as a thousand years, while pressure continued to build.

At 2:46 p.m. on March 11, 2011, about eighty miles offshore, the strain along the tightly squeezed plate surfaces finally became too great. Like a stubborn load winched too tightly, the Pacific Plate broke free and lurched downward with a jerk. That freed the North American Plate to spring upward, the pressure relieved. Along a seam of more than 180 miles, the Pacific Plate dropped at an angle, moving 100 to 130 feet westward as the overlying North American Plate rose up and angled eastward the same distance.

Within moments, the northern half of the island of Honshu, Japan's largest, was stretched more than three feet to the east, a movement of land mass so great the earth's axis shifted by several inches.

Earthquakes send out packets of energy in a successive series of waves. First come the body waves, which travel through the entire body of the earth. The fastest body waves are primary (P) waves, which are nondestructive and travel three or four miles per second. This is the ground motion Japan's earthquake early-warning sensors detect. P waves are followed by secondary (S) waves, which travel at about half the speed but have the potential to cause more damage. Following the P and S body waves are the even slower surface waves, which cause the most severe ground motion and are responsible for the most damage to surface structures. The lead time provided by P waves is crucial because even a few minutes' advance notice allows people to seek safety and critical systems to shut down.

Japan's earthquake warning network is regarded as the best in the world. It was created in the aftermath of the 1995 Kobe earthquake in western Japan, which killed more than five thousand people. The system relies on a thousand ground-motion sensors around the country that can pinpoint the location of an earthquake within a second or two and in most cases estimate its magnitude as fast.

On March 11, sensors near Sendai detected the first offshore tremors within eight seconds and transmitted the information to the Japan Meteorological Agency 190 miles away in Tokyo. Just two days before, four quakes had been measured in this same area of seabed. The largest had a magnitude of 7.3, which was nothing out of the ordinary by Japanese standards.[1] Seismologists thought that the stress had been relieved and the earth had settled back to normal. Now, however, the sensors were recording an even larger quake, with an estimated magnitude of 7.9.

For all of its technological prowess, Japan's early-warning system has a proven weak spot: because of its speed, the system broadcasts warnings before all seismic data have arrived, and thus it tends to significantly underestimate the size of an earthquake. Although a 7.9 magnitude quake could produce major damage, it was not beyond what Japan had often experienced before.

But nature was throwing technology a curveball.

The Fukushima Daiichi nuclear power station in 2010. The exhaust stacks between Units 1 and 2 and between Units 3 and 4 can be used to discharge radioactive gas vented from the containment structure during an emergency. Administrative buildings, including the emergency response center and Seismic Isolation Building, are located behind Unit 1. *Tokyo Electric Power Company*

The shaking continued for about three minutes, an unusually long time. About seventy-five seconds in, long after the system had sent out an alert, the massive undersea plates slid apart, releasing cascading amounts of energy. The U.S. Geological Survey would later estimate that enough surface energy was thrown off by the rupture to power a city the size of Los Angeles for a year. Over several days, the Japan Meteorological Agency upgraded the quake, ultimately to magnitude 9, forty-five times more energetic than the original 7.9 prediction. This was the largest earthquake ever recorded with instruments in Japan and one of the five most powerful in the world since modern record keeping began in 1900.

On a forested stretch of the coast south of Sendai, seismic sensors at Tokyo Electric Power Company's Fukushima Daiichi Nuclear Power Station also registered the P waves. At forty-six seconds past 2:46 p.m., the motion caused a sensor on the Unit 1 reactor to trip. Twelve seconds later, another sensor on Unit 1 tripped. The reactor began to shut down automatically, just as it was designed to do. At 2:47, an alarm alerted control room operators to Unit 1's status: "ALL CR FULL IN."[2] The ninety-seven control rods—the brakes that halt a nuclear chain reaction—were fully inserted into the core. Units 2 and 3 soon followed suit. In about a minute, the three operating reactors at Fukushima Daiichi were in shutdown. (The other three reactors at Fukushima Daiichi, Units 4, 5, and 6, were out of service for routine maintenance.) Elsewhere along the northeastern coast, eight reactors at three other nuclear plants also automatically shut down. It was what U.S. reactor operators call a "vanilla scram"—a shutdown done by the book.

Fukushima Daiichi has six boiling water reactors (BWR), designed in the 1960s and early 1970s by the General Electric Corporation (GE) and marketed around the world. This type of reactor produces electricity by boiling water to make steam to turn a turbine generator. Three factors are at work: the amount of water, its temperature, and its pressure.

As the water circulates around the ferociously hot core, it turns into steam. The steam, under pressure, moves through pipes to the turbine generators. After forcing the turbines around, the steam flows down into a condenser, where it is cooled, converted back to water, and then recirculated through the core. Too little water in the core and the fuel rods can overheat and boil dry; too much steam, and pressure builds to unsafe levels.

The nuclear fuel used in the reactors consists of uranium oxide baked into ceramic pellets. These pellets are formed into fuel rods by inserting them into thirteen-foot-long tubes (or "cladding") made of a metallic alloy containing the element zirconium, which contains radioactive gases produced in the fission process.

The fuel rods are assembled into rectangular boxes known as fuel assemblies, which are then arranged in an array approximately fourteen feet wide within a six-inch-thick steel reactor vessel. The reactor vessel itself is located within another structure called a drywell, which is part of the primary containment structure. In "Mark I" BWRs like Fukushima Daiichi Units 1 through 5, the drywell is made of a steel shell surrounded by steel-reinforced concrete. This shell is a nearly impermeable barrier designed to contain radioactivity in the event of an accident.

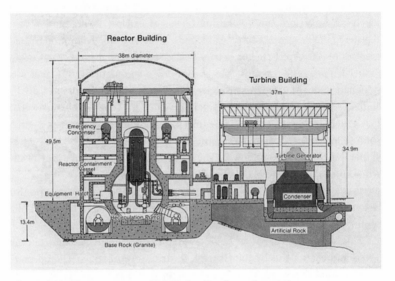

The Mark I boiling water reactor. *U.S. Nuclear Regulatory Commission*

Another part of the primary containment, the wetwell, sits below the drywell and is connected to it through a series of pipes. Half filled with water and often called a torus because of its doughnut shape, this system is designed to reduce pressure by drawing off excess steam and condensing it back to water. Without the "energy sponge" provided by the torus, the primary containment structure would have to be five times larger—and thus more expensive—to withstand the same amount of energy released during an accident.

Finally, the primary containment is surrounded by another structure, the reactor building, known as the "secondary containment." Although the reactor building can help to contain radioactivity, that is not its main purpose.

The ceramic fuel pellets, the zirconium alloy cladding, the reactor vessel, the primary containment, and the reactor building constitute a series of layers that are intended to prevent the release of radioactivity to the environment. However, as the world would soon witness at Fukushima Daiichi, these barriers were no match for certain catastrophic events.

• • •

After a scram, the reactor operators' primary job in the first ten minutes of an abnormal event is to verify that appropriate automatic actions are taking place as the plant shuts itself down. However, with nuclear reactors, the safety challenges continue after the off switch is flipped. Although the chain reaction has stopped and uranium nuclei are no longer undergoing fission, the fuel in the reactor cores continues to generate a huge amount of heat from the decay of fission products, unstable isotopes produced when the reactor was operating. Therefore, pumps driven by electric motors are still needed to circulate cooling water around the nuclear fuel and transfer the heat energy to what engineers call the "ultimate heat sink"—in this case, the Pacific Ocean.

If cooling is lost, in as little as thirty minutes the water level within the reactor vessel drops about fifteen feet and falls below the tops of the nuclear fuel rods. Soon afterward, the exposed rods overheat, swell, and burst. The zirconium alloy cladding reacts with steam and generates potentially explosive hydrogen gas. The temperature of the fuel pellets continues to rise until they begin to melt, emitting more radioactive gases into the reactor vessel. After several hours, the melting core slumps and drops to the bottom of the vessel. The molten fuel is so corrosive that within a few hours it burns completely through the six-inch steel wall.

Once the reactor vessel is breached, the fuel flows through to the concrete floor of the primary containment, where it reacts violently with the concrete, churning out additional gases. At this point, any of several mechanisms can cause the primary containment to fail, either rapidly through violent explosions or slowly through gradual overpressure. The last barrier to the environment—the reactor building—may contain some of the radiation but cannot be counted on to do so. The end result: the release of plumes of radioactive material and contamination of the environment.

In addition to more than two hundred tons of fuel in the cores of Units 1, 2, and 3 combined, the Fukushima Daiichi site stored hundreds of tons of irradiated fuel that had been discharged from the reactor cores and was now being kept under water in swimming pool–like structures at each of the six reactors and in a common pool nearby. Most of this was "spent fuel" no longer useful for generating electricity. Some older spent fuel was stored in "dry" concrete and steel storage casks. Even though the spent fuel rods in the pools had been removed from the reactor cores months or even years earlier, they still generated enough decay heat to require active cooling systems.

Ordinarily, the electricity needed to power the cooling systems for both the scrammed reactors and the spent fuel pools would come from off-site through the power grid. However, conditions on March 11 were far from ordinary because of the earthquake, which had toppled electrical transmission towers and damaged power lines. Inside their control rooms, jolted by the tremors, operators at Fukushima Daiichi could only surmise what was happening in the world beyond. They now watched their monitors as temperature and pressure in the reactors decreased.

In the event that the primary cooling system fails, boiling water reactors have auxiliary systems. Fukushima Daiichi Unit 1 had isolation condensers: large tanks of water designed to provide an outlet for steam from the reactor vessel if it becomes blocked from its normal path to the turbine condenser. The other reactors were each equipped with a "reactor core isolation cooling" system, known as RCIC (pronounced *rick-sea*), which is powered by steam and can run reliably without AC power as long as batteries are available to provide DC power to the indicators and controls.[3] In addition, BWRs are equipped with emergency core cooling systems should the primary and auxiliary systems fail.

Just before 2:48, as the shaking worsened, alarms in the Unit 1 control room signaled that power had been lost to the circuits that connected to the off-site power grid. Like homes and businesses all along the battered coast, the reactor was now without external power. The lights flickered as Unit 1 and the other reactors were automatically transferred from the external power supply to the on-site emergency system. Within seconds, Fukushima Daiichi's thirteen emergency diesel generators (two per reactor, plus a third at Unit 6) automatically fired up. This restored the power, instrumentation, and cooling equipment needed to keep the nuclear fuel from overheating.

At the same time, the turbine generators at Units 1, 2, and 3 shut down, and the valves carrying steam to the turbines automatically closed, as they were supposed to do after a scram. At Unit 1, a rise in pressure was halted five minutes later, at about 2:52 p.m., when other valves automatically opened—also as expected—and allowed the steam to flow into the isolation condensers. About ten minutes later, operators managed to start the RCIC systems at Units 2 and 3.

Once again pressure levels headed downward, as did water temperatures. But just as heat and pressure spikes pose a threat to hardware in a reactor, so do rapidly falling temperatures; both can cause metal to expand or contract too quickly and ultimately break because of high stress. At 3:04 p.m., fearing that the Unit 1 reactor was cooling too fast, control room operators followed procedures and shut down the isolation condensers, figuring they could be turned back on when needed.

• • •

When the tremors had subsided, Fukushima Daiichi workers assembled for a roll call in the parking lot in front of the main office building, which had been damaged by the shaking. Those assigned emergency management duties then moved to the plant's emergency response center on the second floor of the earthquake-proof Seismic Isolation Building next door. From there, they could communicate with operators in the control rooms, who confirmed that Units 1, 2, and 3 had successfully shut down.

But the earthquake was just nature's first assault. The next was about to strike. The quake's hypocenter was located eighteen miles beneath the ocean floor, but the rupture angled upward through the crust, reaching the seabed and reshaping it. That displaced a mountain of water. Some was now heading east into the open ocean. (By the time the waves hit Antarctica, about eight thousand miles south of the epicenter, they still had enough power to break off more than fifty square miles of ice shelf, twice the area of Manhattan.) Waves were also racing westward—headed straight for northeastern Honshu at the speed of a jetliner.

Barely a swell on the ocean's surface, the huge surge of water would slow when it hit shallower depths but then rear up and intensify, striking not once but multiple times. Just as the Japanese are no strangers to earthquakes, they know well the power of the massive wave that they—and the rest of the world—call a *tsunami*.

The northeastern shoulder of Honshu is known as the Tohoku region, a remote mountainous area that includes, from north to south, the prefectures of Aomori, Iwate, Miyagi, and Fukushima. While the region near Fukushima Daiichi is relatively flat, elsewhere steep hillsides and jagged inlets shelter small fishing or farming villages. The portion north of Fukushima is known as the Sanriku Coast and is home to some of Japan's most spectacular scenery and renowned seafood.

For earth scientists, however, the name Sanriku is synonymous with the convulsive forces of nature, as evidenced by a June 1896 earthquake and tsunami, the deadliest in Japan's modern history. A magnitude 7.2 earthquake struck offshore, rattling the coast but causing no alarm. Thirty-five minutes later, at about 8:00 p.m., the ocean suddenly receded hundreds of yards, then returned as a wall of water that destroyed everything in its path. The wave, estimated at 125 feet high in places, killed 22,000 people along the Sanriku Coast and swept away entire villages.[4]

Japan's written records of earthquakes go back to 599 A.D. and document one that struck Sanriku in July 869 A.D., known as the Jogan earthquake. Believed to have had a magnitude of about 8.6, it generated a twenty-six-foot wave that swept inland at least two and a half miles, killing a thousand people, according to an official record.

That was the first of as many as seventy recorded tsunamis that struck the Tohoku region of Japan. The March 2011 tsunami that hit Sanriku and areas to the south, including Fukushima Prefecture, recorded forty-five-foot waves. The Japanese have officially designated this disaster the 2011 Tohoku-oki Earthquake.[5] Its toll: nearly nineteen thousand killed or officially declared missing. Of those deaths, more than 96 percent are attributed to the tsunami.

Fukushima Daiichi's site superintendent, Masao Yoshida, was in the emergency response center when he learned from a television broadcast that the tsunami predictions had been revised upward. Three minutes after the early quake warnings, the coastal prefectures closest to the epicenter—Iwate, Miyagi, and Fukushima—had been alerted to prepare for a "three-meter or higher" wave (ten feet or so). The wave was now estimated at about twice that. Yoshida began to worry that the tsunami might damage emergency seawater pump facilities on the shore—the systems needed to carry residual heat away from the reactors and support equipment, such as the water-cooled emergency diesel generators. However, he expected that even in that case he would be able to compensate by using other available equipment. He could not anticipate the full extent of the disaster about to occur.

At 3:27 p.m., forty-one minutes after the earthquake began, the first tsunami wave hit the seawall extending outward from the Fukushima Daiichi plant. The wave, thirteen feet high, was easily deflected; the wall had been built to withstand water almost thirty-three feet (ten meters) high.

At 3:35 p.m., a second wave struck. This one towered about fifty feet, far higher than anyone had planned for. It destroyed the seawater pumps Yoshida had worried about and smashed through the large shuttered doors of the oceanfront turbine buildings, drowning power panels that distributed electricity to pumps, valves, and other equipment. It surged into the buildings' basements, where most of the emergency backup generators were housed. (Two workers would later be discovered drowned in one of those basements.) Although some diesel generators stood on higher levels and were not flooded, the wave rendered them unusable by damaging electrical distribution systems. All AC power to Units 1 through 5 had been lost. In nuclear parlance, it was a station blackout.

The radioactive waste treatment facility at Fukushima Daiichi is engulfed in seawater at 3:35 p.m. on March 11, 2011, after a wave about fifty feet high slammed into the nuclear power plant. The image below shows the same scene one minute later. *Tokyo Electric Power Company*

Japanese regulators, like their counterparts around the world, had known for decades that a station blackout was one of the most serious events that could occur at a nuclear plant. If AC power were not restored, the plant's backup batteries would eventually become exhausted. Without any power to run the pumps and valves needed to provide a steady flow of cooling water, the radioactive fuel would overheat, the remaining water would boil away, and the core would proceed inexorably toward a meltdown.

Because Japanese authorities, like those in other countries, believed the possibility of such a scenario was very remote, they dragged their heels in addressing this threat. They were confident that the electrical grid and the backup emergency diesel generators were highly reliable and could be fixed quickly if damaged. They refused to consider scenarios that challenged these assumptions.

But in the 1990s, after U.S. officials finally took action to address the risk of station blackout, Japanese regulators also recommended that plant operators develop coping procedures.[6] These plans took advantage of backup cooling systems like the RCIC that operate by steam pressure alone, without the need for electric pumps. But the systems do rely on eight-hour batteries to power equipment that operators can use to monitor their performance and make adjustments to keep them stable. Coping with a station blackout is essentially a race against time to restore AC power before the batteries run down.

Across Japan, a network of remote cameras is mounted at ports, along highways and bridges, atop buildings, and in other critical areas. On the afternoon of March 11, these cameras provided the world with a real-time view of the disaster. Under its agreement with the government, NHK could activate the cameras from its newsroom, giving viewers stunning images: fishing boats tossed like bathtub toys and whole villages consumed by ravaging water, miles of mud, and debris. Video uploaded from personal cell phones and webcams appeared on the Internet within seconds. Billions of people became eyewitnesses to the natural catastrophe unfolding in Japan.

From a hillside behind the reactor complex to which they had evacuated, employees watched as the murky gray water roared up and around the buildings. Cars and machinery bobbed like corks, and debris borne on the surging water crashed into structures like battering rams. Then, with as much force as it had slammed into the plant, the water surged back into the roiling ocean, carrying away equipment and leaving enormous damage behind.

In the capital, officials at Tokyo Electric Power Company, known as TEPCO, assembled their own command center on the second floor of corporate head-quarters. More than two hundred employees, hastily summoned, took their places at long rows of desks.

At 3:37 came the call from Fukushima Daiichi. Not only was Unit 1 without AC power, but it had also lost DC power: flooding had destroyed its backup batteries. The control room for Units 1 and 2 went dark; instrument panels faded to black. All power to these reactors had been lost or couldn't be delivered where needed because of damage to power panels and cables. This was now the most severe type of station blackout. Without DC power to control cooling systems like the isolation condenser, only a very narrow time window was left before core damage began.

One by one, in the space of a few minutes, Units 1 through 5 lost AC power supplies, and high-voltage electrical panels were flooded. (Only one generator, located on higher ground near Unit 6 and cooled by air rather than seawater, continued to work.) For the first time in history, a nuclear accident was unfolding in multiple reactors at the same time.

This was a situation no one had prepared for—or even thought possible. TEPCO's station blackout guidelines were incapable of addressing the challenge now playing out, because they assumed that only one unit would be affected and it could draw on the power supplies of adjacent units. Nor did the guidelines contemplate the simultaneous loss of both AC and DC power. "We encountered a situation that we had never imagined," Yoshida would say later.

To compound the crisis, these reactors had just been subjected to a record-breaking earthquake and tsunami that may have caused structural damage. And now, with power gone, there was no way to know what was happening inside them.

If a natural disaster could trigger a crisis like the one unfolding at Fukushima Daiichi, then, one might wonder, why aren't even more safety features required to prevent such a catastrophic thing from occurring?

The short answer is that developers of nuclear power historically have regarded such severe events as so unlikely that they needn't be factored into a nuclear plant's design. The experts could not imagine that such a cascading failure of safety systems would really occur. So the regulations required only that reactors be able to survive conditions occurring during far less severe accidents, known as "design-basis accidents." As defined by the U.S. Nuclear Regulatory Commission (NRC), design-basis accidents were scenarios that were unlikely to occur during the lifetime of a nuclear reactor but were nonetheless conceivable enough to warrant measures to limit their severity. Reactor designs are equipped with emergency cooling systems and containment structures intended to function during design-basis accidents to limit core damage and radioactive release to a level regulators consider acceptable.

If these systems don't work, then the plant enters the realm of "beyond design-basis" accidents, also called "severe" accidents. Severe accidents challenge plant operators in part because they involve complex and poorly understood phenomena. What is known comes primarily from tests and computer simulations that provide only limited insight. The postmortem review of the 1979 beyond-design-basis loss-of-coolant accident at the Three Mile Island reactor in Pennsylvania confirmed some predictions but raised questions about others. Fukushima Daiichi has raised even more questions, and one day should provide additional answers.

Well before Fukushima, critics argued that predicating reactor safety on the ability to handle design-basis accidents left nuclear plants vulnerable to far worse events that are more probable than the industry would like to believe and than the public would be willing to accept. Even so, design-basis accidents do create a stiff set of requirements. Reactor containments must be rugged enough structures to withstand the high pressures and temperatures that could occur in a design-basis accident without developing large leaks or rupturing. Containments are typically made of steel shells or steel-reinforced concrete with leak-tight steel liners. Because of their size and strength requirements, they don't come cheap.

When GE designed its boiling water reactors, it equipped them with "pressure suppression" containments to reduce construction costs. These containments featured additional systems for converting excess steam to water and thus, the theory went, did not need to be built to withstand very high accident pressures.

After GE began selling the first boiling water reactors with this feature, safety critics called attention to what they believed was a dangerous vulnerability: the complex pressure suppression system was not well understood. If it failed to sufficiently reduce steam pressure during an accident, the too-small primary containment surrounding the reactor vessel might burst, releasing radioactive material into the environment.

And that was not the only threat. The Three Mile Island accident raised awareness of the danger of hydrogen. During that accident, hydrogen produced by the reaction of steam with the fuel rod cladding caused an explosion in the containment. Although the Three Mile Island containment was sufficiently large and robust to withstand the shock, engineers realized that the explosion would have been powerful enough to rupture the smaller, weaker pressure suppression containments. As a result, the NRC required that reactors with pressure suppression containments be retrofitted to control hydrogen accumulation in accidents either by filling the containments with nitrogen, an inert gas, or by activating spark plug–like devices called igniters to gradually burn off any hydrogen. The Japanese were monitoring U.S. developments closely and also required Mark I containments to be "inerted." That addressed one accident contingency, but neither the NRC nor the Japanese worried about what would happen should hydrogen escape the containment and leak into other areas of the plant.

Fukushima Daiichi Unit 1 was a Mark I reactor that began service in March 1971 and ran for forty years. Units 2, 3, 4, and 5 were more advanced BWR models but also had Mark I containments. Unit 6 was a BWR with a Mark II containment, which also used pressure suppression. Across Japan, there were twenty-eight boiling water reactors. The United States has thirty-five, of which twenty-three have the GE Mark I containment. (See p. 279 for a list.)

In Tokyo, Prime Minister Kan was also struggling to grasp what was happening. The capital hadn't been included in the initial earthquake alert because the Japan Meteorological Agency had underestimated the earthquake's size and subsequent hazard zone. But when the shaking hit Tokyo, Kan had hustled out of the hearing room and headed across the street to his office. There, he gathered a small group of advisors in a basement situation room.

Kan had won election to the office of prime minister just ten months before. A member of the Democratic Party of Japan, he had served as finance minister and had campaigned on a promise to improve the nation's weak economy. He also pledged to lessen the influence of the powerful but unaccountable bureaucracy that has long run the government.

When it came to nuclear energy in Japan, that bureaucracy was large. Responsibility was divided among multiple government agencies, whose missions sometimes overlapped—or conflicted. Japan's fifty-four commercial nuclear power plants were regulated by the Nuclear and Industrial Safety Agency (NISA), which operated under the jurisdiction of the Ministry of Economy, Trade and Industry (METI). NISA shared some responsibilities with the Ministry of Education, Culture, Sports, Science and Technology (MEXT), which had a dual role: to promote nuclear energy and to ensure its safe operation. MEXT performed environmental radiation monitoring and assisted local governments with radiation testing in the event of an accident.

Also in the mix was the Nuclear Safety Commission (NSC), an independent agency that operated within the executive branch. The NSC supervised the work of MEXT and METI and provided policy guidance, but also worked to promote nuclear power. And finally, there was the Japan Nuclear Energy Safety Organization, which inspected nuclear facilities, conducted safety reviews, and, in the case of an emergency, made recommendations on evacuations.

Japan's prefectures had a role, too. They were responsible for radiation monitoring and directing evacuations if needed. On paper, all these duties and responsibilities may have seemed clear. In practice, however, the system proved unworkable.

At 3:42 p.m., Tokyo Electric Power Company declared a "first level" emergency, a legal threshold meaning that an accident is predicted or has occurred. By law, TEPCO had to notify the head of the Ministry of Economy, Trade and Industry along with the governor of Fukushima Prefecture and the mayors of the towns of Okuma and Futaba, the communities in which the plant is located. The procedural requirements were clearly spelled out. The notification, according to the plant's emergency plan, was to be done by sending a fax "all at once, within fifteen minutes." (Plant managers were advised to follow up by phone.)

But the emergency plan didn't fit this emergency. There was no power. Phone lines and cellular towers were damaged or destroyed. Faxes or phone calls would be difficult, if not impossible. No one apparently had thought that an event fierce enough to damage a reactor might also disrupt basic communications.

Inside Fukushima Daiichi's emergency response center, a generator powered a video link to TEPCO headquarters. But communications within the plant itself were difficult. The paging system was disabled; TEPCO had provided only one-hour batteries for some of the mobile units and there was no way to recharge them. Crew members often had to return to the emergency center to report simple details—a time-consuming and risky procedure. In many respects, the emergency communication system at Fukushima Daiichi reflected the underlying premise of the plant's comprehensive accident management plan, which read: "The possibility of a severe accident occurring is so small that from an engineering standpoint, it is practically unthinkable." The follies resulting from this complacent attitude began to build catastrophically.

Under the provisions of Japan's Act on Special Measures Concerning Nuclear Emergency Preparedness, regulators from NISA were to staff an off-site command post and help coordinate the emergency response. At Fukushima

Daiichi, the designated center was located about three miles from the reactors. When three NISA workers arrived there, they discovered no power, phone service, food, water, or fuel; additional staff couldn't reach the facility because of damage to roads and massive traffic jams. Equally problematic, the building was not equipped with air filters to protect those inside in case of a radiation release. (The lack of filters had been cited two years earlier by government inspectors, but NISA had failed to install them.) It seemed nobody in government imagined a nuclear accident could produce a cloud of radiation intense enough to pose a hazard a few miles away.

Back in Tokyo, things were not going much better. Kan was in the situation room with his close advisors, few of whom knew much about nuclear power plants. Cell phones didn't work in the basement, making contact with the outside world difficult. Five floors above, the government's nuclear experts had gathered in another emergency response center. Some senior managers from TEPCO joined them. But the two groups were not communicating with each other, despite being in the same building. Nor did anyone from the government head to TEPCO to ascertain what the utility was doing. In many respects, government officials were functioning much like the operators in the control rooms: without information to guide them.

Under normal circumstances, the reactor operators at Fukushima Daiichi had access to a wide range of information about the status of critical systems via the Safety Parameter Display System for each unit. But when the control rooms were disabled by the loss of electrical power, the steady flow of information had largely ceased.

At 3:50 p.m., someone in the shared control room for Units 1 and 2 wrote on a whiteboard about their reactor cores: "Water levels unknown." Without DC power, operators could no longer monitor or manipulate the isolation condensers at Unit 1 or the RCIC at Unit 2 remotely from the control room. Even worse, if those systems were not working and water levels dropped significantly, operators could not start up the emergency core cooling systems at either unit to pump water into the reactors quickly. Things appeared to be a little better at Unit 3. There, control room operators still had some backup battery power that provided readings on pressure and water levels and enabled them to operate steam-driven cooling systems. At about 4:00 p.m., they were able to restart the RCIC system and add water to keep the fuel rods in the Unit 3 core covered.

Units 3 and 4 also shared a control room. With Unit 4 shut down for maintenance, that team focused primarily on Unit 3. The team's colleagues in the Units 1 and 2 control room had their hands full with both reactors, although early on Unit 2 seemed to pose a greater threat because operators could not confirm whether the RCIC was operating or measure the water level in the core. In contrast, the operators believed the isolation condensers in Unit 1 were working, but they could not confirm this either.

To operate the instruments that could provide the information they needed most—the water, temperature, and pressure levels inside the reactors—the engineers at Fukushima Daiichi badly needed power. They thought they had one backup source left: emergency batteries. Soon workers would be roaming the muck- and debris-laden plant grounds, scavenging batteries from undamaged cars and buses in a desperate attempt to jury-rig some sort of power system.

But hooking up the batteries was a challenging task. With much of the plant's electrical infrastructure damaged or destroyed, crews sometimes had to search for working connections behind control room panels or find circuitry elsewhere. Darkness and the presence of standing water made this a delicate and difficult task. Batteries were scarce and too small to provide adequate voltage. The few hours' cushion the operators thought they might get was fast disappearing.

At 4:30 p.m., TEPCO issued a press release announcing that "a big earthquake" had occurred at 2:46 p.m. More than 4 million households were without power. "Due to the earthquake, our power facilities have huge damages, so we are afraid that power supply tonight would run short. We strongly ask our customers to conserve electricity."

The apologetic press release included a reassuring status report on TEPCO's various generating stations in the affected area, including the company's seventeen nuclear reactors: six at Fukushima Daiichi, four at nearby Fukushima Daini,[7] and seven at Kashiwazaki-Kariwa, located on the western coast of Japan. "At present, no radiation leaks have been confirmed," the release noted.

But at Fukushima Daiichi and the utility's command center in Tokyo, the tone was far less confident. At 4:46 p.m., exactly two hours after the first tremors had been detected, TEPCO officially notified the government that the emergency was worsening. Operators could not determine the water level in the reactor cores of Units 1 and 2 and had no assurance that the systems to supply additional water were working. Specifically, emergency core cooling had been lost at Units 1 and 2. This, under law, required the declaration of a "second level" emergency.

The highest priority for the harried team in the darkened Units 1 and 2 control room became restoring water-level indicators. At least then team members might have a better idea of the status of the two units. They salvaged two twelve-volt batteries from buses and additional batteries and electrical cables from a contractor's on-site office.

For a few tantalizing moments, a water-level gauge returned to life, showing that the level was dropping inside the Unit 1 reactor vessel. Minutes later, the gauge died. But this fleeting indication pointed to the possibility that soon water would have to be injected from outside the reactor using portable pumps. A diesel-powered fire pump was started and allowed to idle, ready to inject water into the Unit 1 reactor through a portal normally intended for use in firefighting, not core cooling. In addition to a diesel-driven fire pump at each reactor, there were three fire engines at Fukushima Daiichi that potentially could be used. TEPCO had ordered fire engines deployed to all its reactor sites after a fire broke out at the Kashiwazaki-Kariwa plant following an earthquake in 2007. But the utility did not contemplate that the fire engines might have to be used for something other than firefighting.

Unfortunately, as the plant operators knew, trying to get water into the reactor using either the fire pump or the fire engines would not be easy. These sources could supply water only at a relatively low pressure compared to the pressure within the overheating reactor vessel. Unless operators could depressurize the vessel, they wouldn't be able to force water into it.

As pressure increased inside the reactor vessel, steam flowed out through safety relief valves designed to keep the vessel from rupturing. Pipes leading into the torus carried the steam downward. If the pressure suppression system had been working properly, the steam would have been cooled and turned back into water in the torus, which would reduce pressure throughout the containment. To keep the torus itself from overheating, its water would be routed through tubes into heat exchangers, where seawater flowing around the tubes would absorb the heat and carry it away to the Pacific. But because the seawater pumps were destroyed, there was no effective way to remove heat from the torus.

With no access to the Pacific and no electrical power, there was only one way left to reduce the pressure within the containment: by venting some of the steam into the atmosphere. That would make it easier to inject water into the reactor and would lower the likelihood of a containment breach. Fortunately, the Mark I boiling water reactor was equipped with an emergency vent that, when opened from the control room, would release steam into the environment through a three-hundred-foot-tall stack. As part of the response to a safety review that took place after the 1986 Chernobyl nuclear plant accident, TEPCO had taken measures in the 1990s to improve the effectiveness of the vent system.

One scenario TEPCO had not anticipated, however, was that the vent might be needed during a station blackout, when the valves required to open it could not be operated remotely from the control room. Consequently, the emergency guide did not explain how to operate the valves manually. Nor was the vent equipped with filters to remove radiation from the steam if an emergency release was required. The designers of boiling water reactors believed that filters were unnecessary because radioactive steam would be naturally scrubbed by water in the torus before being vented. But this filtering mechanism had never been demonstrated under real-world conditions, and some experts doubted its effectiveness.

These were two more holes in an emergency plan that was turning out to be full of them. "When on-site workers referred to the severe accident manual, the answers they were looking for were simply not there," investigators would later write. "[T]hey were thrown into the middle of a crisis without the benefit of training or instructions."

Nobody was sure if the ability to relieve the rapid buildup of pressure inside a reactor was within reach. In a BWR accident, venting—a controlled release of radioactivity—is a last-ditch move to stave off a far worse disaster: core melting and failure of the containment, which could result in larger, uncontrolled releases of radioactivity. The amount of radiation released during venting depends on the extent to which the core has been damaged. If it is badly damaged, radiation levels in the steam could be deadly. At this point, no one knew the status of the Unit 1 core, so the relative risks and benefits of venting were not clear. But guided by scant data and instinct, the engineers knew that some sort of intervention to stabilize the reactor was needed—and needed quickly. They moved on two fronts: to bring in additional water supplies and to prepare, somehow, to vent.

Venting, in addition to being technically difficult, was fraught with political and public relations implications; as a result, both TEPCO management and the government in Tokyo would demand a say. Ultimately, however, the decision to vent rested with the most senior official on the scene. At Fukushima Daiichi that was Masao Yoshida, fifty-six years old, who had become the boss ten months earlier.

The superintendent of Fukushima Daiichi, Masao Yoshida. *Tokyo Electric Power Company*

Like a ship's captain, the site superintendent knows the equipment intimately, has firsthand knowledge of the unfolding crisis, and is best positioned to assess the options. Under TEPCO's emergency plan, Yoshida was to make the calls with input from utility executives. At the moment, however, TEPCO's top executives were missing in action.

Chairman Tsunehisa Katsumata was in Beijing on a business trip with Japanese media owners. President Masataka Shimizu had been sightseeing with his wife in western Japan. Shimizu had received an earthquake alert on his cell phone, but his efforts to return to Tokyo that afternoon were thwarted. He had traveled only partway, hoping to make the final leg of the trip home in a TEPCO helicopter. But Japan's civil aviation law bars private helicopters from flying after 7:00 p.m. Late in the evening, he won approval from the government to fly aboard a Japan Air Self-Defense Force (SDF) airplane, but, twenty minutes after the 11:30 p.m. takeoff, the defense minister (unaware of the official authorization) ordered the plane to turn around and instead stand by for disaster relief duties. Shimizu was left on the tarmac. It would be 10:00 a.m. the next day before he and Katsumata made it back to TEPCO headquarters.

Even if TEPCO's executives had been on hand, the decision to vent wasn't solely the utility's to make. Government approval is not required by law, but it was widely understood at TEPCO that government officials from several agencies needed to be brought into the loop. Although no one knew exactly what was taking place inside Fukushima Daiichi, everybody wanted a say. As a result, the decision-making process about venting Unit 1 came to a near standstill.

For the nuclear power establishment, the decision to vent radioactive steam holds serious implications. Releasing radiation into the environment demonstrates unequivocally to the public that this form of generating electricity is not as clean and safe as the industry's public reassurances and promotional campaigns proclaim.

By late afternoon, conditions appeared serious enough that Japanese regulators decided to notify international authorities. Shortly after 4:45 p.m., NISA alerted the International Atomic Energy Agency (IAEA) in Vienna that Fukushima Daiichi had reached the state of "near accident at a nuclear power plant with no safety provisions remaining." Under the IAEA's seven-level scale of nuclear accidents, seven being the most serious, that constituted a level 3 event.

Prime Minister Kan put off declaring a nuclear emergency, despite a request that he do so by the heads of both NISA and METI. The events at Fukushima Daiichi weren't waiting, however. Just before 6:00 p.m., a work crew was sent to the fourth floor of the Unit 1 reactor building, hoping to learn more about the status of the isolation condenser. They wore no protective clothing. As they arrived at the double doors of the reactor building, their dosimeters shot off the scale, and they hurried back to the control room. This strongly indicated that the Unit 1 core was now exposed and fuel rods had already ruptured. Full-scale melting of the core would soon begin.

But the official word from TEPCO was vague and outdated. At 5:50 p.m., the utility issued a press release announcing the "malfunction" of the diesel generators and the resulting loss of backup power more than two hours earlier. "There have been no confirmed radioactivity impact to [the] external environment," the English-language version of the announcement said. "Further details are in the process of being confirmed."

Across Japan, the scope of the natural disaster that had hit the Tohoku coast was still sinking in. More than one hundred thousand members of the Self-Defense Forces, the Japanese equivalent of the National Guard, had mobilized; local disaster agencies were struggling to grasp where to focus their attention; in some communities, whole neighborhoods had simply vanished, their residents washed out to sea. In places along the coast, tsunami survivors were still trapped atop buildings where they had fled the oncoming water. It was cold and darkness was falling. The human toll was staggering. As Kan would later say, "The focus was on saving lives."

Japan's prime minister Naoto Kan announces the emergency response efforts on March 11. After declaring that the nuclear power plants in the region had automatically shut down, he said, "At present, we have no reports of any radioactive materials . . . affecting the surrounding areas." *Cabinet Secretariat, Government of Japan*

That rescue effort took on an added dimension shortly after 7:00 p.m., when Kan declared the nuclear emergency requested two hours earlier. At 7:45, chief cabinet secretary Yukio Edano alerted the nation and the world that an emergency had been declared. "Let me repeat that there is no radiation leak, nor will there be a leak," Edano said in a reassuring voice. The prime minister's office apparently was unaware of the high readings taken earlier at the Unit 1 reactor building. That news certainly wasn't coming from TEPCO. At 9:00 p.m., the company issued another press release warning of a possible power shortage.

Shortly afterward, Yoshida got what he thought was a reprieve: the water gauge on Unit 1 suddenly started working, indicating the water level was still almost eight inches above the top of the fuel. (In all likelihood, the gauge was providing an inaccurate reading because it was not calibrated for extreme conditions.) Not long after receiving that bit of apparent good news, however, he got the bad news. The radiation levels inside the Unit 1 reactor building had risen so high that entry was forbidden, seriously complicating any emergency repairs. The radiation readings were positive proof that the fuel core now was exposed and most likely melting. That finding was passed on to Tokyo along with Yoshida's alarming prediction that the Unit 2 water level and RCIC status were unknown and that the fuel there could also be uncovered soon.

By then, authorities had ordered the emergency evacuation of those living within about a two-mile (three-kilometer) radius of the reactors. Officials of the towns of Okuma and Futaba dispatched sound trucks and local firefighters to go door-to-door with the announcement. Many of the residents were still reeling from the earthquake and tsunami, searching for missing loved ones, or scavenging for their possessions. They were told to leave immediately. Those living a little farther out, between two and six miles (three to ten kilometers) from the reactor, were directed to stay indoors. All they were told was that there were problems at Fukushima Daiichi.

As people fled, the first of about a dozen power supply trucks were rumbling toward Fukushima Daiichi, dispatched that afternoon from TEPCO headquarters and from other utilities. These mobile generating units might provide the power so desperately needed. But the challenge of even getting to the plant was enormous. The drivers were forced to navigate roadways battered by the natural disaster and clogged with traffic leaving the area. By 11:00 p.m. the first trucks had made it, and workers attempted to connect the generators but had difficulty locating functioning electrical power panels. Unfortunately, some of the cables were too short and the plugs incompatible.[8] False alarms of another tsunami interrupted the task, forcing workers to flee to higher ground. After twenty-four hours, only one generator was operating.

During the first hours of the accident, as crews at Fukushima Daiichi scrambled, government and utility officials in Tokyo lacked a similar sense of urgency, a response some later attributed to a failure of leaders there to understand what was happening at the plant. They seemed to feel that there was adequate time to decide a course of action. Prime Minister Kan, however, became increasingly frustrated that the venting was not happening.

The delay in venting was the result of a number of factors. One was the difficulty of the emergency evacuation. There was an informal agreement with the government of Fukushima Prefecture that until nearby residents were safely relocated, venting would not occur. But even if workers had tried to start the venting immediately, they would have had problems. Because the vent valves could not be operated from the control room, they had to be opened manually. It took hours to figure out where the valves were physically located and which could be opened by hand.

By the time the evacuation was declared complete and the decision to vent was finally reached on the morning of March 12, conditions at the plant had worsened. Accessing and opening the vent valves located deep inside the dark, intensely hot, and now radioactive reactor building was a far more dangerous mission than it would have been the previous evening. This was a scenario no accident drill had covered.

In Tokyo, Kan's irritation over the slow flow of information—and its accuracy—was mounting. In addition to the two hundred TEPCO technical advisors on duty at company headquarters, four hundred plant personnel under Yoshida's direction manned Fukushima Daiichi's emergency response center. Communication between headquarters and the response center was occurring via TEPCO's in-house videoconferencing system.

The government, on the other hand, had no similar ability to communicate with the plant. In Tokyo, NISA obtained information from phone conversations with TEPCO. (A videoconferencing system was not set up until March 31, when the government and TEPCO created a joint response center.) Dissatisfied, Kan and his advisors asked TEPCO to assign staff members to the prime minister's office for briefings, and Kan and his aides eventually even began calling Yoshida for answers, an action for which Kan would be later accused of micromanaging.

Just before midnight on March 12, the plant's emergency team was able to get a pressure reading of the Unit 1 drywell using a portable generator in the control room. The team found that it exceeded the design maximum operating pressure. At 12:49 a.m. on March 12, Yoshida decided the pressure in Unit 1 was likely so high that venting now must take place. TEPCO president Shimizu, who still hadn't returned from his vacation, agreed with the decision at about 1:30 a.m. But TEPCO also wanted the government's blessing. At this point, an unanswered question was whether TEPCO might have to vent Unit 2 as well as Unit 1.

Unit 1 was the only reactor of the six that relied on isolation condensers for emergency cooling. Apparently, even the shift team assigned to Unit 1 was unfamiliar with that design. Had team members been trained in the system, they would have recognized that it was not operating and that Unit 1 had been deprived of water for hours.

Meanwhile, the situation at Unit 2 also remained a mystery. Although the RCIC system had started up at the time of the earthquake, once DC power to the control panels was lost no one could tell whether the RCIC had continued to successfully inject water into Unit 2. Yoshida feared it hadn't and that the top of the fuel was about to be exposed.

A crew wearing breathing gear and protective clothing ventured into the RCIC room in the basement of the Unit 2 reactor building to determine its status. The first trip was inconclusive. A second team was dispatched. This crew said it believed the RCIC was functioning based on pressure measurements. With that news, Yoshida decided that Unit 1 warranted first priority; Unit 2 could wait. Word was sent to Tokyo. Kan and Banri Kaieda, head of the Ministry of Economy, Trade and Industry, agreed.

TEPCO managing director Akio Komori, who had once worked at Fukushima Daiichi, joined Kaieda and the head of NISA at a joint press conference shortly after 3:00 a.m. to announce the venting. If it was meant to provide a reassuring message, it fell far short. Just before the press conference began, the three men found they had differing information about the status of the reactors. Uncertain of what was actually occurring at Units 1 and 2, they decided that Komori would announce the venting but not identify the reactor involved. When questioned by the media, he became confused. A few minutes later, government spokesman Yukio Edano took the podium to say that radiation would be released but the public should remain calm.

Edano was getting his information from NISA, which was getting it from TEPCO. When later asked to assess the accuracy of information coming from his office to the Japanese public at this time, Kan would say that NISA officials were "choosing their words carefully," and as a result Edano was being misled.

The Unit 1 drywell, at twice its design pressure, was likely approaching a failure point. Venting had to happen, and it had to happen now. Engineers at the plant hurriedly calculated possible radiation exposure from the release. As the preparations continued, a worker was sent to check radiation levels at the Unit 1 reactor building. When he opened the door, he saw "white smoke" inside and quickly left without taking a reading. The smoke, whatever it was, clearly showed that something was leaking somewhere. At about 4:00 a.m., radiation levels near the plant's main gate were measured at 0.0069 millirem (0.069 microsieverts) per hour. Twenty minutes later, they had jumped nearly tenfold to 0.059 millirem (0.59 microsieverts) per hour. The Unit 1 drywell was now venting itself.

RADIATION AND THE BODY:
DOSES, DAMAGE, AND DEBATE

Radioactive materials emit ionizing radiation—that is, radiation energetic enough to detach electrons from atoms, turning them into charged particles (called ions). Exposure to ionizing radiation can have different effects on the human body, depending on the extent and nature of the damage it causes on the cellular level.

The relative biological damage in the human body resulting from radiation exposure is measured in units called sieverts. The United States, unlike most of the rest of the world, uses a rem ("radiation exposure man") as its standard measure. One sievert is equivalent to 100 rem.

One class of radiation effects is known as "deterministic," meaning that a certain level of exposure will almost always cause a particular outcome. Deterministic effects generally result from levels of radiation high enough to kill cells, causing widespread damage to tissues or organs. Depending on the nature of the injury, such doses range from tens to hundreds of rem delivered over a short period. The resulting injuries include burns, cataracts, thyroid nodules, hair loss, gastrointestinal distress, low blood counts, and cardiovascular disease. Recent studies have also identified statistically significant excess risks of certain circulatory diseases at low doses. These studies suggest that the mortality from such diseases due to low-dose radiation exposure may be comparable to that from cancer.

As the dose increases or wider areas of the body are exposed, the victim may develop an illness known as "acute radiation syndrome." Although this sometimes can be cured with treatment, high enough doses—above several hundred rem delivered in a brief period—will almost certainly result in death within days or weeks. Following the Chernobyl accident, twenty-eight people—plant workers and firemen in close proximity to the damaged reactor—are known to have died in this manner. Death from acute radiation syndrome would be classified as an "early fatality," occurring within a few weeks or months after exposure to a nuclear plant release.

Deterministic effects feature a dose "threshold" below which a particular effect will not occur. This is because cells must sustain a certain amount of damage before the cell dies. In addition, a certain number of cells must be affected before enough tissue damage occurs to cause clinical symptoms.

The other major class of radiation effects is known as "stochastic," or random. Ionizing radiation can cause DNA damage that *might* produce changes in cellular behavior leading to cancer, but does not *definitely* cause such changes. Cancer risk does rise with increasing doses, however, because the more DNA lesions there are, the higher the chance that one of them will lead to cancer.

It is currently estimated, based on studies of survivors of the atomic bombings at

Hiroshima and Nagasaki, that a dose of ten rem delivered at once will raise an individual's lifetime risk of fatal cancer by about 1 percent on average. This risk is higher for children and other groups of people (for instance, those with certain genetic variations) who are more sensitive to the effects of radiation than the average adult. Doses delivered over lengthy periods may be less effective at causing cancer, but there is much uncertainty about whether this is true. Because most cancers take many years, even decades, to appear after exposure to radiation, deaths due to stochastic effects are called "latent cancer fatalities."

Increases in cancer risk associated with low doses of radiation are small compared to the background level of cancer in humans. Thus, in epidemiological studies it is hard to detect a direct cause-and-effect relationship between radiation in the low-dose range (generally below about five rem) and cancer. However, there is broad scientific consensus that all radiation exposure, no matter how small, produces some increased risk of cancer. It is biologically plausible that even a single particle of ionizing radiation could cause enough damage in a cell to induce cancer.

Nonetheless, a small and vocal group, including some scientists, believes that the absence of observable evidence for an increase in cancer risk at low doses implies that ionizing radiation is harmless below some threshold. The logic is that at these doses, the rate of damage is so low that DNA repair mechanisms can combat it successfully. Some even believe in the theory of "hormesis," which holds that low doses of radiation are actually beneficial and can stimulate the immune system like a vaccine.

Although these arguments have been reviewed and largely rejected by authoritative scientific bodies, such as the National Academy of Sciences' Biological Effects of Ionizing Radiation (BEIR) VII Committee, advocates of a threshold continue to cite them as justification for the belief that the harmful effects of radiation in general, and nuclear energy in particular, are exaggerated.

At the extremes of pressure and temperature that the drywell was now experiencing, the bolts and seals used to make it leak-tight were giving way, allowing radioactive steam and hydrogen gas to escape directly into the reactor building. This was the last barrier preventing the release of radioactivity into the atmosphere. At least the pressure inside the containment was dropping, although not enough to obviate the need for operator-controlled venting. The plant operators needed to reduce the containment pressure by venting through the torus instead of the drywell; then some of the radioactivity might be filtered as the gases passed through water. Otherwise, unfiltered releases directly from the drywell would continue.

To vent the Unit 1 containment, workers needed to manually open two valves in different locations. The first was on the second floor of the reactor building. Without electric power, workers would have to open it with a hand crank—provided they could reach it. The second valve was in the basement torus room. Ordinarily the basement valve required compressed air to operate. By poring over plant drawings, however, workers had located a wheel handle on the valve that they could use to open it manually.

They now mapped out the route they would take to get to it. Their path would be through the dark and hot reactor building. In the past few hours, the plant site had been rattled by twenty-one aftershocks, raising fears of another tsunami. What if another wall of water were to sweep in and surround them all?

But the more immediate threat was invisible: the rising radiation level. Crews now had to wear protective gear and breathing equipment if they ventured out of the emergency response center. Soon, radiation levels rose even inside the Units 1 and 2 control room, and operators there had to wear full face masks and protective clothing. Most of the control room team moved to the Unit 2 side of the room and crouched down on the floor, where the radiation was a little lower, to do their work.

Workers wearing protective clothing and respirators inside the Unit 2 control room on March 26, 2011, after power is restored at the plant. *Tokyo Electric Power Company*

Meanwhile, workers continued their efforts to force water into the Unit 1 core by any means available. The diesel-driven fire pump idled for hours, but it remained useless because operators could not depressurize the reactor vessel and the pump was not powerful enough to inject any water into it. Finally, the pump shut down and could not be restarted.

And then, an apparent miracle happened, but it was a mixed blessing. At around 2:00 a.m., operators were able to recover some instrumentation using batteries and discovered that the reactor vessel pressure had dropped significantly all by itself. On one hand, the pressure, although still high, was now low enough to give the fire engines on-site a chance of getting water to the core. On the other hand, the depressurization was an ominous sign that the reactor vessel had sprung a leak somewhere—although it is not clear that anyone appreciated that at the time.

At about 4:00 a.m., workers finally managed to connect a fire hose to a portal on the Unit 1 turbine building, providing a pathway for one of the fire engines to pump freshwater into the reactor core. It took almost two more hours to establish a consistent flow rate, but for the first time in nearly fifteen hours, water was reaching the core. However, it was too little, too late. The fire engine pump pressure was still too low to force much water into the vessel. And in any event, it is likely that by then the core had already melted through the bottom of the reactor vessel and dropped to the containment floor. If so, much of the water being injected into the core was probably flowing out into the containment.

The whole exercise had taken much longer than anyone had anticipated in part because no plant worker knew how to operate the fire engines maintained for emergencies at the reactor site. A contractor had to be convinced to help with the arduous task. With rising radiation levels, ongoing aftershocks, and the threat of another tsunami, the contractor was reluctant to agree.

In Tokyo, at both the utility and the prime minister's office, everyone was asking the same questions: When was the venting going to begin? Why was it taking so long? When Prime Minister Kan asked the TEPCO official assigned to his office to explain the delay, the only answer he got was: "I don't know the reason."

Shortly after 6:00 a.m., Kan decided to find out for himself. In the midst of the growing crisis at the plant, Yoshida was informed that the prime minister was en route via helicopter. Before Kan left Tokyo, he ordered authorities to widen the evacuation zone around the plant from about two miles (three kilometers) to six miles (ten kilometers). Without careful coordination, however, that would put a lot of people on the roads when the venting finally did take place. Nonetheless, while Kan was airborne, METI minister Banri Kaieda ordered TEPCO to vent.

At 7:11 a.m., Kan, accompanied by Haruki Madarame, the chairman of the Nuclear Safety Commission, landed at Fukushima Daiichi. Yoshida explained the difficulties to Kan, who calmed a bit when he discovered that he and Yoshida had attended the same college, the Tokyo Institute of Technology. Kan repeated the order: vent. Yoshida promised that would happen by about 9:00 a.m. Fifty minutes after he arrived, Kan headed back to Tokyo and Yoshida back to managing the disaster. Although Kan's dramatic fly-in garnered much publicity (and eventually was portrayed by his critics as dangerous meddling), Yoshida remained the man calling the shots.

Three two-person teams suited up with protective clothing. They were about to enter a dark, highly radioactive reactor building that they might have to flee at any moment because of the intermittent earthquake tremors. They knew they would have just a few moments to accomplish their task before they would reach their allowable radiation dose and have to retreat. The makeup of the teams reflected the radiation exposure risks everybody recognized: young employees were excluded from the mission.

At 9:02 a.m., the plant was notified that the evacuation of residents was complete and that the venting could now begin. (The information about the evacuation turned out to be incorrect.) The first team entered the second floor of the Unit 1 reactor building, flashlights providing the only illumination as team members searched for and located the hand crank. With tools about the size of those used to change a car tire, they cranked the vent valve open a quarter of the way before they hastily left. In the ten or so minutes they had been inside, each man had received a radiation dose of two and a half rem (twenty-five millisieverts), one-fourth of the total dose they were permitted to receive under emergency conditions for an entire year. When members of the second team entered the torus room in the basement of Unit 1 to open the second valve, they found dose rates so high that they were unable to reach it, and they fled. Even so, one of the operators exceeded the emergency dose limit of ten rem (one hundred millisieverts). Entry efforts were then abandoned.

A new plan was devised; it, too, proved problematic. There was another, larger valve that might be opened from the control room with battery power and a compressed air supply. But the available compressor didn't work without electricity, and nobody had a portable unit. Once again, contractors' offices on and off the plant site were searched. More time elapsed.

Workers then attempted to open the small air-operated valve in the torus room from the control room in the hope that there was still enough residual air in the valve to make it work. For venting to be successful, this valve would have to remain open long enough to create pressure in the vent sufficient to burst a rupture disk, which served as the last barrier between the radioactive gas and the environment. When radiation levels spiked at the main gate and at several monitoring posts at about 10:40 a.m., operators thought their venting attempt had worked and the disc had ruptured. However, soon afterward radiation levels began to fall and it was no longer clear that there had been significant venting.

Finally, shortly after noon, a portable compressor was located and jury-rigged so it could be connected to the system that normally supplied air to plant equipment. Together with temporary DC power, this allowed operators to open the large air-operated valve from the relative safety of the control room. At 2:00 p.m., the compressor was started and the large valve opened. Pressure dropped inside the containment, and presumably inside the reactor vessel as well, allowing operators to inject water at a higher rate.

Another sign that the operators might have succeeded came when NHK cameras trained on the plant from a distance captured white smoke emerging from the exhaust stack shared by Units 1 and 2 and rising high above the sky-blue reactor building. Yoshida and the exhausted crew in the emergency response center thought they had finally caught a break. At 3:18 p.m., he notified TEPCO in Tokyo of the venting.

Captured by an NHK camera positioned about twenty miles away, the explosion inside Unit 1 at 3:36 p.m. on March 12 destroys the roof of the reactor building and blows out a panel in the adjacent Unit 2 reactor building. *NHK*

Yoshida had other progress to report as well. Recognizing that freshwater supplies were limited, he had ordered workers to come up with a method for utilizing seawater to cool the reactors. After freshwater supplies were exhausted shortly before 3:00 p.m., he gave the order to prepare to use the Pacific. It was a complicated endeavor—workers had to position three fire engines with interconnecting hoses to pump seawater transferred into a pit—but by 3:30 p.m., the lines to inject seawater into the Unit 1 reactor were almost in place. A steady supply of vital cooling water for the core was now within reach.

And then, at 3:36 p.m.—almost exactly twenty-four hours since the tsunami had roared in and flooded the plant—a powerful explosion ripped through the Unit 1 reactor building, blasting off the roof and sending debris everywhere. The impact blew out a panel in the neighboring Unit 2 reactor building. Inside the emergency response center, stunned workers watched the explosion on television, not sure exactly what had blown up.

2

MARCH 12, 2011:
"THIS MAY GET REALLY UGLY . . ."

At 9:46 a.m. on March 11 local time, the U.S. Nuclear Regulatory Commission officially entered what it calls Monitoring Mode, a heightened state of readiness to respond when a nuclear incident is unfolding. It had been exactly nine hours since the earthquake tremors were first detected at Fukushima Daiichi.

NRC personnel with a wide range of expertise assembled in the agency's Operations Center, a low-ceilinged room crowded with desks and computer monitors. The NRC is currently housed in three modern mid- and high-rise office buildings in Rockville, Maryland, just outside Washington, DC. The headquarters complex is often simply called White Flint, after a nearby Metro stop.

Overnight, news media had begun carrying stories about the natural disaster in Japan, where local time was fourteen hours ahead of Washington. Now, details about mounting problems at the Fukushima Daiichi reactors vied with accounts about the devastation from the earthquake and tsunami. The NRC staff was scrambling to learn more.

Earlier that morning, the National Oceanic and Atmospheric Administration had issued a tsunami warning for the U.S. Pacific coast, triggering concern about a possible threat to the Diablo Canyon Power Station in California and other coastal nuclear facilities. (When the wave did arrive at about 8:30 a.m. local time, it amounted to a surge no larger than the normal tides; U.S. nuclear operations were not affected.) The information available from Japan at that time, however, convinced NRC officials that the disaster there had now risen far beyond any West Coast consequences. Responsibility shifted from the NRC's western regional office to headquarters. Ultimately more than four hundred staff members at White Flint and around the country would be drawn in.

The Operations Center at the Nuclear Regulatory Commission (NRC) headquarters outside Washington, D.C. More than four hundred NRC staff members nationwide were involved in analyzing the accident at Fukushima Daiichi. *U.S. Nuclear Regulatory Commission*

Details were arriving from a variety of sources; most of the news was secondhand. Japan's *Asahi Shimbun* was reporting that a state of emergency had been declared and evacuations were planned. CNN reported that radiation levels were increasing. On the other hand, Reuters quoted the World Nuclear Association in London, an industry trade group, as stating that the situation in Japan was "under control."

Among NRC staff members across the agency, e-mails were flying furiously as they combed the Web and TV networks for information, sharing their finds with each other and struggling to put the sometimes conflicting pieces together. Soon, however, those pieces appeared to add up to an ominous accident, as their e-mails reveal:

9:21 a.m.: "We are safe and lucky this time."

9:42 a.m.: "[M]y understanding is that this is a Station Blackout and if they don't get some kind of power back it's only a matter of time before core damage. This is a really big deal."

More details were shared—and analyzed from a distance of nearly seven thousand miles. Within minutes of the early morning press briefings in Tokyo announcing plans to vent radiation, the implications of that decision were obvious at White Flint.

3:07 p.m.: "This may get really ugly in the next few days."

Even as they struggled to comprehend the toll on human lives and property of the natural catastrophe that had struck Japan, the men and women of the NRC knew that the disaster unfolding inside the reactors half a world away had implications for White Flint. The fate of Fukushima Daiichi could dramatically affect the commission's own future, which is closely tied to the fortunes of the nuclear industry.

Among the messages crisscrossing the NRC as the crisis deepened was one that put this concern in a nutshell:

"Is this safety bad or economics bad?" asked an e-mail sent to Brian Wagner of the NRC's Office of New Reactors.

"Both," Wagner replied.

Like many regulatory agencies, the NRC occupies uneasy ground between the need to guard public safety and the pressure from the industry it regulates to get off its back. When push comes to shove in that balancing act, the nuclear industry knows it can count on a sympathetic hearing in Congress; with millions of customers, the nation's nuclear utilities are an influential lobbying group.

Over the years since its establishment in 1974, the NRC has been accused by many critics of favoring the industry's point of view when it comes to adopting higher safety standards. To the critics, when the question was "safety" versus "economics," it seemed that economics often won. The NRC's supporters responded that nuclear power generation was already so safe that tighter requirements were rarely worth their cost to industry.

The crisis now beginning in Japan plainly had the potential to undercut the safety argument. What the NRC staffers could not yet know was how clearly Fukushima Daiichi would demonstrate the dangers that arise when regulators become too close to the industry they oversee.

Japan is not the United States; the relationships between government and business in the two nations are as different as other aspects of their cultures. But the history of nuclear power in Japan, and the incestuous practices that fostered it, do provide dramatic evidence that giving the nuclear industry the benefit of the doubt can lead to unimaginably dire consequences.

·   ·   ·

At the moment an NRC staffer speculated things "may get really ugly," it was 5:07 a.m. March 12 in Japan. Inside the Seismic Isolation Building at Fuku-shima Daiichi, Masao Yoshida was struggling to figure out how to vent the Unit 1 reactor. Prime Minister Kan, frustrated with a lack of information, was about to head to the plant himself. And thousands of Japanese embarked on the first of what would be a torturous series of migrations as they evacuated from a two-mile radius around the plant.

This wasn't the first time Japanese citizens had had to flee a runaway nuclear accident. Nor was it the first time they had had reason to question the response of nuclear authorities.

In September 1999, inside a nondescript industrial building in Tokaimura, a village about seventy miles northeast of Tokyo, three employees of the JCO Company were preparing a small batch of 18.8-percent-enriched uranium fuel—containing a far higher concentration of uranium-235 than the typical fuel used in power reactors—for use in an experimental fast neutron reactor. The workers were untrained in handling material of this enrichment level and were preparing the fuel in stainless steel buckets. In a hurry, they poured the solution into a tank, bypassing safety controls. The mixture went critical, initiating a self-sustaining chain reaction that cycled on and off for hours, periodically emitting high levels of gamma and neutron radiation. The in-dustrial building provided little radiation shielding because under normal conditions there was no need for it.

Government response to the accident was slow, even as radiation spread nearly a mile away. Almost three hours elapsed before a radiological monitor-ing team from the national Science and Technology Agency was dispatched to the scene; the team members initially did not believe there was much reason for concern. Five hours after the incident began, about 160 residents were ordered out of their homes. Seven hours after that, about 310,000 people were told to stay indoors for twenty-four hours. Workers piled sandbags and other shielding materials around the facility to reduce dose rates outside the fence. It took twenty hours to halt the criticality. The three workers at the plant were severely overexposed; two died.

Reaction to what was at the time Japan's worst nuclear accident was muted. Two-thirds of Tokaimura residents surveyed said they were now critical of nuclear power. But about half saw their village's future as "co-existing with the nuclear industry," not surprising since a third of the populace worked at one of the dozen or more nuclear facilities nearby. (This part of Japan is known as "nuclear alley.")

Outside Japan, however, the events in Tokaimura and Tokyo's response came under sharp attack. A week after the accident, a scathing editorial in the British journal *Nature* placed responsibility "squarely on the shoulders of government," specifically, its Science and Technology Agency, "which is proving itself incapable of adequately regulating the safety of nuclear power." (A government reorganization in 2001 shifted regulatory responsibility to a new body, the Nuclear and Industrial Safety Agency, or NISA, which would be overseen by the Nuclear Safety Commission.)

"The Japanese government seems unable to set up competent regulatory bodies with sufficient staff and expertise," the *Nature* editorial continued. The NSC "is a group of part-time academic experts who rubber-stamp documents produced by a small team of officials, who are far too few in number and lack the expertise needed to regulate the safety of such a huge and potentially dangerous industry.

"Will the situation improve significantly after this accident? Based on the record to date, probably not."

Before the end of 1999, Japanese lawmakers passed the Nuclear Emergency Preparedness Act, intended to improve cooperation among various levels of government in the event of an accident. The law also established an elaborate notification and response framework. Sadly, however, *Nature*'s pessimistic assessment was on the mark. The modest reforms implemented after Tokaimura did not forestall the confusion and lack of preparedness that helped make the Fukushima accident so much worse than it had to be.

While the Cold War helped drive support for the nuclear industry in the United States, the industry's privileged status in Japan has much deeper economic roots. Japan is a nation that consumes large amounts of energy but has few resources of its own; it currently imports all but 16 percent of its energy needs. Only China, the United States, India, and Russia consume more energy than Japan, and they are much larger countries.

Achieving energy self-reliance has long been a major preoccupation for the Japanese. As the nation struggled toward recovery after World War II, energy security became the lynchpin to its comeback. Power shortages during the Korean War, coupled with soaring demand from industry and new residential customers, led the Japanese government to step in with assistance for the country's utilities, of which TEPCO, the Tokyo Electric Power Company, was the oldest and largest. "Through direct and indirect means, the power companies reaped enormous sums from Treasury coffers," according to scholar Laura E. Hein. Once the utilities, which held regional monopolies, had improved their solvency and invested in additional fossil and hydro plants, Japanese authorities turned their attention toward atomic power.

In Japan, radiation has a heightened symbolism. In 1945, the United States dropped atomic bombs on Hiroshima and Nagasaki, resulting in an estimated 140,000 deaths and leading to Japan's surrender in World War II. Despite this indelible legacy, the prospect of deriving abundant energy from the atom was too enticing to ignore. On March 4, 1954, the Diet approved a budget for nuclear energy development.

The United States stood ready to help. Cold War tensions were increasing; the Soviet Union was becoming a global threat, and the United States was eager to cement an alliance in the Pacific. Washington's Atoms for Peace plan became the ideal vehicle. The Eisenhower administration and those who favored private nuclear development in the United States were looking for ways to promote peaceful uses at home and abroad.

In September 1954, Thomas E. Murray, a member of the U.S. Atomic Energy Commission (AEC), delivered a speech to three thousand members of the United Steelworkers of America. His topic was the development of nuclear power, but he was just as concerned with flag waving. Optimism was already in the air. Just five days earlier, Lewis L. Strauss, chairman of the AEC, had delivered a similarly upbeat message to the National Association of Science Writers in which he uttered an oft-quoted (later oft-derided) prediction: "It is not too much to expect that our children will enjoy in their homes electrical energy too cheap to meter."

Murray, in his address, spoke eloquently about nuclear power and America's stature. One way to assert and maintain its global leadership, he said, was for the United States to build a nuclear reactor in Japan before the Soviets could.

A reactor would be a "lasting monument to our technology and our goodwill," Murray told the union members. "Because the economics of nuclear power are so uncertain, it is unrealistic to expect private industry to undertake on a purely risk basis anything like the effort that the world atomic power problem demands." Shying away from nuclear power for cost reasons alone would be "inconsistent with all this nation stands for" and would "play into the hands of the Soviets."

This came as great news, certainly for the steelworkers, who would reap skilled manufacturing jobs, but also for U.S. companies like Westinghouse Electric Corporation and GE, which already had a toehold in Japan. The previous year the two manufacturers had loaned Japan $6.1 million to help the government purchase thermal generating equipment. Now, with the U.S.

government proselytizing its benefits, they saw an opportunity to market their newest technology: nuclear energy. For the Japanese, the atom held out the promise of finally attaining a secure energy future and fueling growth. In August 1956, construction began on the country's first nuclear facility, a research center, in Tokaimura. The village soon would also become home to Japan's first nuclear reactor.

The appeal of nuclear power in Japan went beyond central government bureaucrats. Where plants were built, local economies benefited in ways that strengthened the nuclear industry's political as well as economic clout.

In 1958, the governor of Fukushima Prefecture approached TEPCO, hoping to persuade the utility to build its first reactors along an underdeveloped stretch of coastline. The motive was purely economic. Rural areas of Japan, like this region and the Sanriku Coast to the north, were losing population to cities. Eventually, a portion of the coast shared by the towns of Futaba and Okuma was chosen for the new plant.

Local officials welcomed the project, but kept their negotiations with TEPCO secret, fearing some residents might not be so enthusiastic. TEPCO, for its part, dispatched young female employees to accompany utility engineers inspecting the proposed site, both disguised as vacationing hikers to avoid arousing public suspicions. Only two years after a deal was sealed did local residents finally learn of the construction plans.

Ground was broken for the first reactor at Fukushima Daiichi in July 1967. TEPCO chose GE as the contractor, a vendor it had used for conventional power plants. GE would build one of its new designs, the Mark I boiling water reactor. (The first two Mark I reactors, at Oyster Creek in New Jersey and Nine Mile Point in New York, began generating power in 1969, while the Fukushima plant was under construction.) TEPCO, which by 1970 was the world's largest privately owned utility, would stick with GE designs for the five other reactors eventually built at Fukushima Daiichi and for two of four units at nearby Fukushima Daini.

Local officials' predictions that the giant Fukushima Daiichi complex would deliver a financial bonanza proved true. In an effort to promote nuclear projects, the national government provided subsidies to local governments. As the reactors were constructed and came on line, local property tax revenues soared. By 1978, the town of Okuma derived nearly 90 percent of its tax revenues from the plant, which also provided badly needed jobs. Residents enjoyed sports facilities and other amenities. Many other reactor projects were launched along Japan's coasts to take advantage of seawater for removing waste heat.

Japan's nuclear quest entwined government and private industry in a tight alliance. The push by business leaders, politicians, academics, and bureaucrats steamrolled the few experts who raised warnings. One such expert was a young seismologist, Katsuhiko Ishibashi. A newly minted PhD from the prestigious University of Tokyo, Ishibashi discovered that a fault line west of Tokyo was much larger than previously assumed. The Hamaoka nuclear power plant sat atop that fault. Ishibashi's research findings were published in 1976, the same year Hamaoka's Unit 1 began producing electricity. By then, Japan had twenty reactors operating or under construction. Not until two years later, in 1978, did Japanese authorities draft seismic guidelines for reactor design.

Finding an area in Japan not susceptible to seismic activity was difficult. Advances in seismology were revealing the existence of heretofore unidentified faults that had produced massive earthquakes in the past, and most likely would produce them in the future. The embrace of nuclear power by government and industry was not about to be slowed, however. Time and again, utilities and regulators downplayed or ignored the threat posed by earthquakes.

For Ishibashi, the issue became a lifelong crusade. In 1997, he coined the term *genpatsu-shinsai*, a catastrophe involving a quake-induced nuclear accident. He envisioned a scenario in which there was a loss of power to a reactor and "multiple defense systems lose their function simultaneously," resulting in the release of radiation over a wide area.

Most members of the Japanese public paid little attention to such warnings; for them, the development of nuclear power went hand in hand with an improving economy. Japan boasts one of the most reliable electrical delivery systems in the world. Its ten utilities promoted reactors as a way to ensure that economic growth—and the lifestyle it provided—would continue.

Not everyone was happy. In 2003, residents living near Hamaoka—which by now had four operating reactors—sued to shut them down, arguing that they were unsafe and could not withstand a major earthquake. Testifying on behalf of plant owner Chubu Electric was Haruki Madarame, a University of Tokyo professor and nuclear proponent, who would eventually be appointed chairman of Japan's Nuclear Safety Commission, a position he held on March 11, 2011. Madarame had publicly scoffed at the warnings of Ishibashi

and the Hamaoka plaintiffs—including the possibility of a simultaneous failure of emergency generators—saying such concerns would "make it impossible to ever build anything." A court concluded that the safety measures at
Hamaoka were adequate.

Other lawsuits challenging reactor safety also met with defeat. In 1979,
residents in the area of Kashiwazaki, a town in Niigata Prefecture on the Sea
of Japan, asked the courts to overturn a license granted to TEPCO to build
what would become the world's largest nuclear plant, with seven reactors. The
lawsuit claimed that the government had failed to perform adequate inspections of the geology of the plant site and had overlooked an active fault line.
The lawsuit wound on for a quarter century, while the Kashiwazaki-Kariwa
nuclear plant was built and began operating; then, in 2005, a court ruled there
was no fault line. Two years later, a magnitude 6.8 quake struck off the coast
ten miles from Kashiwazaki-Kariwa. A fire broke out at the plant, which was
designed to withstand only quakes of magnitude 6.5 or lower.

"What happened to the Kashiwazaki-Kariwa nuclear plant should not be
described as 'unexpected,'" Ishibashi wrote in the *International Herald Tri-
bune* shortly afterward. This was the third major earthquake to strike near a
Japanese nuclear facility in two years. But as the plant suffered no significant
damage from the quake, regulators remained confident that existing seismic
standards were adequate.

Other legal challenges on safety issues routinely failed. According to an
analysis by the *New York Times* after the 2011 earthquake, fourteen major
lawsuits raising reactor safety questions had been filed against the Japanese
government or utilities since the late 1970s. Evidence often revealed that
operators had downplayed seismic hazards. In only two instances did courts
rule for the plaintiffs, and those decisions were overturned by higher courts,
the *Times* reported. "If Japan had faced up to the dangers earlier," Ishibashi
told the *Times* soon after the March 2011 disaster, "we could have prevented
Fukushima."

That's not to say the Japanese authorities simply dismissed the earthquake
threat. In 1978, parliament passed the Large-Scale Earthquake Countermeasures Act, a law based on the belief that quakes could be predicted well in
advance.

The quest to forecast earthquakes is as old as the science of seismology
itself. "Ever since seismology has been studied, one of the chief aims of its students has been to discover some means which would enable them to foretell

the coming of an earthquake," wrote John Milne in 1880. Milne, an English geologist and mining engineer who became interested in earthquakes while teaching in Japan, helped found the Seismological Society of Japan that same year. He is considered the father of modern seismology in part because of his work in that country.

Accurate predictions, however, have remained a tantalizing but elusive quest. "Journalists and the general public rush to any suggestion of earthquake prediction like hogs toward a full trough," said C.F. Richter, from whom the scale measuring earthquakes' energy got its name, when accepting the Medal of the Seismological Society of America in 1977. "There is nothing wrong with aiming toward prediction, if that is done with common sense, proper use of correct information, and an understanding of the inherent difficulties."

Richter's admonition notwithstanding, many came to believe they could not only predict earthquakes with accuracy but also determine which areas were likely to be safe from serious damage and therefore suitable for nuclear reactors and other potentially hazardous facilities.

Debates over site suitability or the adequacy of standards ignored the inescapable fact that no one could claim true knowledge of the massive forces always at work far below the topography of Japan. No matter how rigorous the standards applied to nuclear plant siting and construction, there would always be uncertainty; and it wasn't clear if science could say how large those uncertainties actually were.

That reality didn't deter the adherents of prediction or those who believed that limited standards could ensure public safety. Among those promoting this belief were the nuclear industry and those charged with regulating it.

Japan's new seismic protection law was directed toward a single event: a shallow magnitude 8.0 earthquake in the Tokai region, about sixty-two miles (one hundred kilometers) southwest of Tokyo. Seismologists such as Kiyoo Mogi of the University of Tokyo had long warned of a disastrous quake in the densely populated region. (While the legislation was pending, a magnitude 7.0 earthquake struck nearby, causing twenty-five deaths and widespread property damage.) The act set up a network of monitoring stations intended to give three days' advance warning. Soon, the public—and government— came to assume that the next Big One in Japan would hit Tokai. And they'd get plenty of notice.

Mogi cautioned that the underlying concept of the law—that there would be an unmistakable warning—was flawed. Earthquakes aren't predictable in the same way tides or sunsets are, he wrote in 2004; instead, scientists must base judgments on complex data whose relation to seismic events is not fully understood. Mogi was troubled that the government was so confident in its ability to predict earthquakes that it continued to license nuclear plants in an area of high risk: namely, Hamaoka.

Japanese officials had begun to let the numbers take on a life of their own. The probabilities—themselves subject to debate—began to be viewed as accurate predictions, and they contributed to the overconfidence that would be shattered in March 2011.

Japan later added another layer to its earthquake prediction system. In the aftermath of the deadly 1995 Kobe earthquake, the government established the Headquarters for Earthquake Research Promotion. In 2005, the first National Seismic Hazard Maps of Japan were published. The maps are based on surveys of active faults, long-term estimates of the probability of earthquake occurrence, and evaluations of strong ground motion. They reflect a belief that "characteristic earthquakes" occur at predictable intervals.

The concept of hazard mapping provokes controversy among seismologists and emergency planners around the globe. Some geophysicists argue that accurate hazard maps are impossible to produce. Rather than providing reliable information, they say, the maps tend to create a false sense of security. Others, including the U.S. Geological Survey, support hazard mapping, arguing that although the maps are not perfect, they do offer some guidance for purposes such as setting building codes.

The maps are based on centuries' worth of data about earth movements. Even so, some of the largest earthquakes in recent years, even before March 11, 2011, occurred in areas or with a degree of force that surprised many scientists: the Indian Ocean (2004), China (2008), New Zealand (2010 and 2011).

With expert views ranging from faith that earthquakes could be predicted with certainty to disbelief that they could be predicted at all, confusion reigned in the field of seismic risk. Apparently even state-of-the-art science was unable to shed much light on the question of "how safe is safe enough" when it came to building nuclear power plants in Japan.

The nuclear accidents at Three Mile Island in 1979 and Chernobyl in 1986 stalled nuclear development in many places. Japan wasn't one of them. There, construction continued full throttle. During the five years after Chernobyl, Japan added five new reactors, with several more under construction.

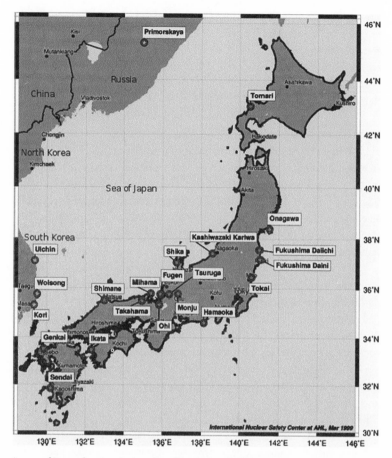

A map of Japan showing the location of its nuclear facilities. *International Nuclear Safety Center, U.S. Department of Energy*

The accident at Three Mile Island Unit 2 resulted from a series of equipment malfunctions, design flaws, and operator errors. Eventually about half of the fuel core melted. Nearly 150,000 people fled their homes. In the end, radiation releases from the reactor were small, but the accident symbolized to the world that nuclear power was not the safe form of energy generation its backers claimed.

Seven years after the accident at Three Mile Island, engineers at the Soviet-designed Chernobyl Nuclear Power Plant in the Ukraine were conducting a safety test on the Unit 4 reactor. After a sudden power surge, operators attempted an emergency shutdown, but power spiked instead, rupturing the reactor core and setting off a series of explosions. The graphite used to moderate fission in the core ignited, sending a massive plume of radiation over large areas of the Soviet Union and Europe. Ultimately, more than 350,000 people evacuated and resettled; a large area around the plant has been declared uninhabitable.

The Japanese government and news media portrayed Chernobyl as a man-made disaster caused by poorly trained operators and the structural defects of old, Soviet-built, Soviet-maintained equipment. The message was clear: such an event could not happen in Japan. But during the country's own nuclear disaster in 2011, Japan's *Yomiuri Shimbun* noted, in understated fashion: "[T]his assessment was optimistic." "We have to recognize," the newspaper declared in an editorial in April, "that there is no perfect technology"—a difficult admission in a country that prided itself on sophisticated engineering and exacting standards.

From the beginning of the era of nuclear power, the Japanese public had been repeatedly assured by government regulators, plant owners, and the media that it was inherently safe. Eventually, this view was generally accepted, despite the periodic eruption of scandals revealing shortcomings in the competence and integrity of those in charge of nuclear power production.

Headlines scattered over the decades built a disturbing picture. Reactor owners falsified reports. Regulators failed to scrutinize safety claims. Nuclear boosters dominated safety panels. Rules were buried for years in endless committee reviews. "Independent" experts were financially beholden to the nuclear industry for jobs or research funding. "Public" meetings were padded with industry shills posing as ordinary citizens. Between 2005 and 2009, as local officials sponsored a series of meetings to gauge constituents' views on nuclear power development in their communities, NISA encouraged the operators of five nuclear plants to send employees to the sessions, posing as members of the public, to sing the praises of nuclear technology.

The utilities and regulators used the same tactic in the summer of 2011 as local governments debated allowing the restart of reactors shut down after the Fukushima Daiichi accident. Utility employees were encouraged to send anonymous e-mails supporting a restart or to show up at meetings and deliver the message in person. When the collusion was revealed, public opinion turned hostile.

Through it all, the nuclear industry in Japan remained largely unchallenged, insulated by official reassurances that the nation's elaborate oversight system was functioning and the industry's overall performance was beyond reproach, despite a few bad apples.

Challenge was unlikely to come from within the government. Nuclear energy in Japan has been described as "national policy run by the private sector." Regulators and the regulated cohabit peacefully in a "nuclear village" whose mission is to promote nuclear power. When it comes to who is watching over whom, the lines often are blurred. A regulator today can become a utility employee tomorrow and vice versa.

It's known as a "revolving door" in the United States. In Japan, moving from a public to a private sector job is called *amakudari*, "descent from heaven." [1] Bureaucrats know that when they are ready to retire, often at an early age, a comfortable job could be waiting in the industry they once regulated. There's little incentive to rock the boat.

Shortly after the Fukushima accident, Japan's *Yomiuri Shimbun* reported that thirteen former officials of government agencies that regulate energy companies were currently working for TEPCO or other power firms.

Another practice, known as *amaagari*, "ascent to heaven," spins the revolving door in the opposite direction. Here, the nuclear industry sends retired nuclear utility officials to government agencies overseeing the nuclear industry. Again, ferreting out safety problems is not a high priority.

The ties between government and industry go beyond sharing personnel. Japan's *Asahi Shimbun* reported in early 2012 that twenty-two of eighty-four members of the Nuclear Safety Commission and two of its five commissioners had received a total of about $1.1 million in donations from the nuclear industry over a five-year period ending in fiscal 2010. One-third of the NSC members on committees overseeing nuclear operations had received such donations. Prior to becoming chairman of the NSC in 2010, Haruki Madarame, then a professor at the University of Tokyo, had received about $49,000. Critics of nuclear power rarely were on the receiving end of such largesse. The "nuclear village" looked after its own.

On December 26, 2004, a magnitude 9.2 earthquake struck off the western coast of Sumatra, triggering a devastating tsunami across the Indian Ocean that killed more than 286,000 people in fourteen countries. The quake, like the one that struck Fukushima seven years later, was a subduction earthquake. And, like the 2011 quake, it caught seismologists by surprise. When the earthquake hit, a subcommittee of the NSC was in the midst of a long overdue review of Japan's seismic standards for nuclear plants. The review, initiated after the 1995 Kobe earthquake, had been inching along for eleven years.

A major issue before the subcommittee was how to determine the maximum earthquake plants had to be able to withstand. The old standard required that every plant be able to survive a nearby magnitude 6.5 earthquake, which in Japan was fairly common. The new standard would replace that generic criterion with limits tailored to the seismic risks at each plant site. The utilities were concerned that the revised standard could effectively increase the earthquake magnitude their plants would have to withstand and require them to make costly seismic retrofits.

When the new guidelines were ultimately approved in September 2006, critics called them too vague. Among the opponents was Katsuhiko Ishibashi, a subcommittee member, who said the guidelines were full of loopholes. Angry at the subcommittee's unwillingness to consider more rigorous standards, he resigned.

As Japan's largest utility, TEPCO reaped enormous benefits from the blurred lines between government and industry. The company's operating record raised plenty of warning signs about the dangers of such incestuous practices, however. And they were apparent well before March 2011.

In 2000, Kei Sugaoka, a nuclear inspector working for GE at Fukushima Daiichi, noticed a crack in a reactor's steam dryer, which extracts excess moisture to prevent harm to the turbine. TEPCO directed Sugaoka to cover up the evidence. Eventually, Sugaoka notified government regulators of the problem. They ordered TEPCO to handle the matter on its own. Sugaoka was fired.

While TEPCO was ostensibly dealing with the matter, Fukushima Daiichi continued to operate. Then, in late summer 2002, the company admitted it had been falsifying safety records for years, covering up evidence of cracks in thirteen reactor core shrouds at all three of its plants. The shrouds are stainless steel cylinders that hold fuel assemblies in place and help direct the flow of cooling water. At a press conference announcing the cover-up, government regulators declared that public safety was not threatened. They had reached that determination not from their own inspections, but from TEPCO's assurances.

When the cover-up became public, TEPCO's chairman and president resigned. The new chairman declared the falsification of records "the gravest crisis since the company was established." Despite the scandal, both former executives were retained as advisors to TEPCO. Theirs weren't the last heads to roll at the company.

•   •   •

In January 2007, TEPCO admitted to more falsified records, involving a total of about two hundred incidents dating back many years. (A newspaper headline the next day read: "Not Again!") It was a year of catharsis for Japan's nuclear industry. Six other utilities revealed their own unreported safety problems. NISA had directed the utilities to own up to any such issues, hoping to head off public opposition to new reactor construction. Whether the outpouring of misdeeds helped or hurt NISA's cause is not clear.

Among the incidents TEPCO divulged were problems at its Kashiwazaki-Kariwa reactors on Japan's west coast. Kashiwazaki-Kariwa is the world's largest nuclear generating station, capable of producing 8,212 megawatts of electricity, nearly double Fukushima Daiichi's output of 4,696 megawatts. In one instance, during a routine government inspection at Kashiwazaki-Kariwa, operators discovered that a component of the emergency core cooling system wasn't working. Workers made adjustments in the control room to make it appear the pump was functioning.

According to TEPCO's explanation, the utility wasn't the problem; the rules were. "[F]alsification occurred because passing the inspections became the objective," the company told the government. TEPCO admitted only to handling data "inappropriately," not to lying to inspectors. For example, the utility failed to report that control rods came loose more than once in the reactor cores at Fukushima Daiichi, in one instance triggering a criticality accident that lasted seven and a half hours. Technically, TEPCO acted legitimately. Reporting the criticality mishap, for instance, was not required under Japan's safety rules at the time.

YES, THE PAPERWORK MATTERS—A LOT

Fudging figures, doctoring reports, and creative accounting may seem minor, though deplorable, offenses. But they are not minor. When it comes to nuclear plant safety, the value of accurate, complete, and reliable paperwork cannot be underestimated.

In the United States, NRC inspectors audit only about 5 percent of the activities at nuclear plants, according to senior managers at the commission. Most of these audits involve reviewing the records of tests and inspections performed by plant workers. The NRC inspectors themselves witness only a very small fraction of actual tests and inspections. If safety inspectors could not trust a plant's paperwork, they would have to personally observe many more activities than they do now to gain confidence that their assessment of the plant's safety was a reflection of reality. To put this another way: when workers feel free to prepare fictional accounts of tests and inspections, nuclear safety assurances begin morphing from nonfiction to fiction as well.

Then came the earthquake that damaged Kashiwazaki-Kariwa. It occurred on the morning of July 16, 2007, with a rupture along the fault that a court had ruled in 2005 didn't exist. The plant had been built to withstand a smaller quake; this one, with a magnitude of 6.8, created ground motion two and a half times greater than that plant had been designed for. The episode raised questions about TEPCO's response to such crises that would prove all too relevant in 2011.

When the quake struck, three of the seven reactors were at full power; a fourth was in the process of starting up. The rest were out of service for refueling and maintenance. All four operating reactors automatically shut down. A fire broke out in an electrical transformer. Because it was a national holiday, the plant was shorthanded. Emergency response crews were difficult to assemble. The quake had damaged the plant's own fire extinguishing system. City firefighters were pulled away from local emergencies and sent to the plant; their arrival was delayed by an hour because of quake damage.

At the time, Prime Minister Shinzo Abe criticized TEPCO for being too slow in reporting problems at Kashiwazaki-Kariwa. "Nuclear power can only operate with the public's trust," he told reporters. The International Atomic Energy Agency sent inspectors to the shaken plant and concluded that the damage was "less than expected." The agency did, however, recommend a reevaluation of the seismic situation at Kashiwazaki-Kariwa, especially the existence of active faults beneath the site. Later in 2007, TEPCO reported a fourteen-mile-long (twenty-three kilometers) active fault in the seabed eleven miles from Kashiwazaki-Kariwa. The company said it had known about the fault since 2003, but did not report its findings then because TEPCO staff didn't believe the fault would produce an earthquake large enough to threaten the reactors.

Immediately after the 2007 earthquake, the utility took all seven of the Kashiwazaki-Kariwa reactors out of service to check for damage and upgrade their seismic resistance.[2] Without 20 percent of its generating capacity, TEPCO posted its first loss in twenty-eight years, totaling $1.44 billion. Its stock value dropped 30 percent.

To bolster confidence in the wake of the Kashiwazaki-Kariwa debacle, a new TEPCO president was named in 2008. Masataka Shimizu took over, replacing Tsunehisa Katsumata, who became chairman. The front office shuffle was getting almost routine: Katsumata had taken the top job in the wake of the 2002 falsification scandal; now Shimizu, a forty-year veteran of the company, stepped in. Both men had spent their entire careers at TEPCO. Shimizu made cost cutting a high priority, and within two years had returned TEPCO to the black, exceeding his target of $615 million in cuts. His secret? Reducing the frequency of inspections.

Direct structural damage is not the only danger that earthquakes can pose to Japan's coastal nuclear plants. They can also cause devastating tsunamis, as became obvious in March 2011. Yet predicting tsunamis is a similarly inexact science. Given the uncertainties, developing standards for tsunami protection posed another conundrum for regulators and plant owners.

On tsunami protection, TEPCO took its cue from government, and Tokyo seemed in no hurry. The lengthy safety review that led to the 2006 earthquake guidelines did not address tsunamis. Proposed tsunami guidelines were moving through their own separate review process at a pace almost as slow as the tempo at which the quake standards had progressed. The nuclear industry also had a strong presence in these deliberations.

Just as the science of seismology had evolved in the years since Japan's first nuclear plants were built, tsunami research, a newer field, was gaining ground rapidly. And Japan also provided an ideal laboratory: the Sanriku Coast alone had experienced four destructive tsunamis in a little more than a century, including the 1896 quake and wave that had a run-up height of nearly 125 feet.

Those tsunamis had struck to the north of Fukushima Daiichi. When construction began on the reactors in the late 1960s, engineers dismissed the likelihood that the plant location might be vulnerable. Based on the worst historical tsunami on record at the Fukushima site—resulting from a 1960 earthquake in Chile—the reactors were designed to withstand a tsunami with a maximum height of about ten feet (3.1 meters). TEPCO was so confident of this data point that the company actually lowered the height of the bluff where the plant was to be built by more than eighty feet (twenty-five meters). That made it easier to deliver heavy equipment to the site and to pump cooling water into the reactors. The company also said the excavation would enhance earthquake protection by placing the reactors on bedrock—albeit bedrock far below the natural elevation.

So confident was TEPCO that in 2001 it submitted to NISA a single-page tsunami plan that ruled out the possibility of a large tsunami hitting the plant and causing damage. The company provided no data to support its conclusions, and NISA apparently asked for none. "This is all we saw," a NISA official told the Associated Press, which located the TEPCO document a decade after it was submitted. "We did not look into the validity of the content."

NISA lacked authority to question the accuracy of TEPCO's presented data in any event, because tsunami plans from utilities were voluntary at this stage. TEPCO provided its information in advance of new tsunami guidelines being developed by the Japan Society of Civil Engineers. That group called for protective measures to be based not only on historical tsunami data, but also on wave heights calculated by numerical models, taking uncertainties into account. In the one-page plan for Fukushima Daiichi, TEPCO provided its projection using this method: a wave no higher than nineteen feet (5.8 meters), generated by a magnitude 8.0 quake modeled after one that had occurred in 1938. Accordingly, the utility made some modifications, including raising the motors of the seawater intake pumps. (In 2009, TEPCO further refined the calculations and raised its estimate of the maximum tsunami height to about twenty feet.)

The prevailing wisdom, based on the historical record, was that earthquakes larger than magnitude 8.0 would not occur in the offshore trench near Fukushima, although they had occurred further north off the Sanriku Coast. However, in 2002, the Headquarters for Earthquake Research Promotion said it was possible that earthquakes similar to the 1896 Meiji Sanriku earthquake, which had an estimated magnitude of 8.3, could occur anywhere along the trench as far south as the Bousou peninsula, well below Fukushima Prefecture.

TEPCO paid no attention to that prediction until 2008, when experts warned that the utility's tsunami assessments could have underestimated the potential size of earthquakes off the Fukushima coast. As a result, TEPCO conducted a calculation assuming that an earthquake comparable to Meiji Sanriku occurred in that area. Based on this model, the utility now predicted a tsunami up to thirty-four feet (10.2 meters) high near the plant's seawater intake pumps. A wave that large could sweep inland and reach a run-up height of more than fifty-one feet (15.7 meters) around Units 1 through 4 at Fukushima Daiichi. But TEPCO did not consider these results realistic, maintaining that the undersea faults in the area were not the type capable of causing large tsunami-generating earthquakes.

Reaching even further back in seismic time, however, one potential counterexample stood out: the 869 A.D. Jogan earthquake and tsunami. The exact origin of the earthquake, believed to have been a magnitude 8.6, was unknown, but scientists had found evidence of geologic deposits from the tsunami well inland in areas not that far north of Fukushima Daiichi. Could such a devastating earthquake occur close to Fukushima after all? Concerned, TEPCO ran another calculation assuming an earthquake as large as Jogan and found it might produce a tsunami as high as thirty feet. The utility then surveyed tsunami deposits around Fukushima Prefecture and located some just north of the Daiichi site, but it found their patterns to be inconsistent with its model. TEPCO's conclusion: further research was needed.

A tugboat that was swept inland by the March 11 tsunami and left to rest in the devastated town of Ofunato, north of Sendai. Similar scenes of destruction could be found along the battered coastline of northeastern Japan. *U.S. Navy*

TEPCO managers discussed the new wave projections internally and considered countermeasures. But ultimately the company did not see a need to prepare for what it still regarded as a highly improbable event. It didn't move its backup diesel generators out of the turbine building basements. Nor did it enhance protection by constructing a large seawall. Although the Sanriku Coast ranks as one of the most "engineered" in the world—with tsunami barriers stretching for miles—TEPCO management feared that a tall barricade in front of a nuclear plant would send the wrong message to the public. "Building embankments as tsunami countermeasures may end up sacrificing nearby villages for the sake of protecting nuclear power stations," according to a TEPCO document. "It may not be socially acceptable."

NISA conducted hearings on earthquake and tsunami hazards at nuclear plants in June 2009. The panel examining Fukushima Daiichi didn't include a tsunami expert. Earthquakes were regarded as a more probable threat, and that's what the committee focused on. But when the panel's findings were presented, a respected seismologist warned NISA that a tsunami at the plant could be as devastating as an earthquake. When he asked why the 869 A.D. Jogan earthquake, which sent water more than two miles inland, was not incorporated into the panel's assessment, a TEPCO official dismissed it as a "historic" event not relevant to the deliberations. Although NISA promised to follow up on the issue, at the next meeting Fukushima Daiichi's existing preparations were deemed sufficient. Without pressure from regulators, TEPCO continued to investigate the Jogan issue on a slow track. TEPCO did eventually get around to reporting new tsunami damage assessments to NISA—on March 7, 2011.

Four days later, at the U.S. NRC, the prediction that things could get "really ugly" at Fukushima Daiichi was coming true. Inside the Operations Center at White Flint, experts watched the crisis develop with growing alarm—and frustration at the lack of information or the ability to respond. The NRC's early offer of engineering assistance to Japan had been met with silence, and the NRC staff, like ordinary members of the public, had access to nothing but spotty and confusing media reports. "This is not our event," NRC chairman Gregory Jaczko warned his impatient team.

But that didn't preclude the agency's experts from parsing every detail trickling out of Japan. In the early hours of March 12, Daniel Dorman, a deputy director working in the Operations Center, called Jaczko at home to alert him to the explosion at Unit 1. Dorman and his colleagues were watching it replayed on a television screen.

"It's an initial short duration pulse," Dorman told his boss, "like an explosion, followed by a large cloud, and then there is some subsequent footage showing what appears to be the frames of the building that—the upper walls around the—what would be the metal framework above the refueling level, it's been opened up to the I beams." He continued, "We're still working off of what we got on the media. But it is a very disturbing image."

Dorman wasn't alone in his concern.

At 5:12 a.m. Washington time, the NRC's Robert Hardies e-mailed his colleague Matthew Mitchell: "My dog woke me up to go out. I turned on CNN. They had breaking video they could not explain. To me it looked like a containment building disappearing in an explosive cloud. WTF."

3

MARCH 12 THROUGH 14, 2011: "WHAT THE HELL IS GOING ON?"

At first, plant boss Masao Yoshida thought the violent motion beneath his feet at 3:36 p.m. March 12 was another earthquake. Then came potentially worse news: the top of the Unit 1 reactor building had blown off; only the steel framework remained.

Four minutes later, the harried team in the plant's emergency response center joined the rest of the world watching the explosion replayed on TV screens. A huge billow of white smoke and debris expanded skyward, drifting away on the prevailing winds.

A check of the jury-rigged gauge showed that the water level inside the Unit 1 reactor had not changed: 5.5 feet below the top of the fuel, or about 21.5 feet below normal. The fuel had started melting about twenty-one hours earlier. Also, the pressure within the reactor vessel had not dropped, suggesting that the worst-case scenario—that the explosion had damaged the reactor vessel itself—had not occurred.

This looked to personnel at the plant and at TEPCO's emergency operations center like a hydrogen explosion. But from what source? Prime Minister Kan had asked the Nuclear Safety Commission chief, Haruki Madarame, that morning about the likelihood of hydrogen igniting, and had been assured that this was unlikely.

During normal reactor operations, nitrogen gas is added to the drywell—part of the primary containment structure—to preclude an explosion of hydrogen should any accident occur. Hydrogen cannot ignite unless oxygen is present, so substituting nitrogen for oxygen—a process called "inerting"—eliminates the possibility of a volatile mix forming inside the drywell. The atmosphere in the reactor building itself was not inerted because regulators never thought it would be necessary; protecting the containment from rupture would be sufficient to prevent radiation leaks. Now, somehow, hydrogen gas had apparently migrated to an area where no one had expected it to be.

The initial reaction among the startled workers was denial that the hydrogen was generated from damaged nuclear fuel. Yoshida first thought the explosion might involve the turbine generators, which use hydrogen gas as a cooling medium in their operation. However, this was ruled out when reports indicated that the Unit 1 turbine building had not been damaged in the blast.

Unit 1 after a hydrogen explosion blew off the top of the reactor building. *Tokyo Electric Power Company*

Another theory suggested that the spent fuel pool in the "attic" of the reactor building was boiling and the fuel there was no longer completely immersed. If so, the overheated cladding of the fuel rods could have reacted with steam to generate the hydrogen gas that ignited. Without working instruments, there was no way to check the water level or temperature of the pool. But it was not clear how so much of it could have boiled away in such a short time, given what the operators knew about the amount of heat generated by the relatively small quantity of fuel in the pool: only 292 spent fuel assemblies and 100 new assemblies. (The new assemblies were there to await the next refueling.)

Then another hypothesis was put forward: the extreme pressure and heat that had built up before venting had loosened bolts and seals in the reactor's drywell, allowing radioactive steam and hydrogen to leak into the reactor building, where a spark or static electricity triggered the blast. Or perhaps when the containment was vented, hydrogen under enormous pressure had leaked out of the vent pipe into the reactor building before it could be expelled through the tall exhaust stack shared with Unit 2. Normally, ventilation fans within the reactor building might have helped to draw any hydrogen up and out of the building, but the fans couldn't work without electricity. Instead, the gas would have accumulated until it became concentrated enough to explode. No matter what the hypothesis, though, one thing was clear: if the containment had been vented sooner to reduce its pressure, this might have prevented hydrogen—and fission products—from leaking into the reactor building.

For the hapless crew in the plant's emergency response center, there was more bad news: falling debris had damaged the electric cable that workers had spent hours laying and were within minutes of energizing to restore power to Units 1 and 2. The explosion had also damaged the fire hoses that workers had carefully arranged to inject seawater into Unit 1. Although radioactive litter was now scattered around the Unit 1 reactor building, Yoshida ordered his employees to get the hoses working. The fire engines injecting freshwater into Unit 1 had used up their supply nearly an hour earlier. The need to get water into the core was still imperative, and the workers had been very close to being able to supply it. The explosion set them back by several hours, but despite the hazardous conditions they repaired the damage by 7:00 p.m.

Meanwhile, just beyond the six-mile (ten-kilometer) evacuation zone, residents of the coastal town of Namie thought they had finally reached safety. Earlier that morning they had been rousted from the evacuation centers to which they had fled to escape the tsunami; now those centers were too close to a new danger: Fukushima Daiichi. Their next haven was a school. Children played in the schoolyard. Adults, dressed in heavy winter coats, sat together on the gymnasium floor or cooked a meal outside. While the world watched images of the giant cloud billowing from Unit 1, those closest to it were unaware because of the loss of cell phone and television service.

As the Namie evacuees would later learn, merely being farther from the damaged reactors would not necessarily protect them. Luck was on their side that night, however, because the emissions from Unit 1 were relatively small, and the prevailing winds blew them primarily northward along the coast and then out to sea.

Almost from the beginning of the accident, an early-warning system had been making ominous predictions about the danger that the Namie evacuees and other people near Fukushima Daiichi could face should conditions worsen at the plant. The System for Prediction of Environmental Emergency Dose Information, or SPEEDI, was developed by the Japanese after the Three Mile Island accident. It went into operation in 1986. If radiation is released in an accident, SPEEDI uses real-time measurements from the nuclear plant together with meteorological data to predict where the radioactivity will spread and how intense it will be. When actual radiation release data are available, the system is far more useful in assessing risks and defining evacuation areas than the concentric zones arbitrarily drawn around a plant. The system is continually updated and monitored from an office in the capital.

The day before, when TEPCO declared a nuclear emergency, SPEEDI had switched to emergency mode, ready to track any releases from Fukushima Daiichi and guide decisions about evacuations based on the location of radiation plumes. There was one problem: Fukushima Daiichi had no power. As a result, when radiation was released, the plant's on-site measuring devices were unable to gather and transmit data about it to SPEEDI. Without such data, all SPEEDI could do was indicate the direction in which the radiation would travel.[1]

Those projections were duly passed from the Education Ministry, where SPEEDI is housed, to NISA, which forwarded them to the prime minister's office. But they came with a caveat: their reliability could not be guaranteed because no one knew how much radiation was actually being released. As a result, the readouts were never shown to the prime minister.

Four days into the accident, the news media began asking for the SPEEDI data. That prompted a meeting of top Education Ministry officials who worried that "a release of the predictions could cause people unnecessary confusion." With that, responsibility for SPEEDI data was handed off to the NSC, whose chief, Madarame, dismissed the calculations as no better than "a mere weather report."

The Namie evacuees at the school had no inkling they were at risk until a TEPCO employee showed up late on the afternoon of March 12 in protective clothing, carrying a dosimeter. Less than an hour after the Unit 1 explosion, radiation readings at the plant boundary had soared. But the employee reassured the worried families that they were safe because they were outside the official evacuation zone—and then he jumped in his car and left. About 6:30 p.m., military trucks arrived at the school. With little official explanation, the occupants were made to pack up their few belongings and head out into the night, once again in search of safety. Other evacuees weren't as lucky; some remained in areas that were in the path of radiation plumes until March 16 before being told to leave.

In Tokyo the evening of March 12, officials gathered in the prime minister's office, struggling to figure out what was happening at Fukushima Daiichi. Just before 6:30 p.m., a decision was made to double the evacuation zone to a 12.4-mile (twenty-kilometer) radius around the plant. Kan announced the decision on national television, telling his audience that Japan was facing an "unprecedented crisis." Joining him was Cabinet Secretary Edano, who reassured the public that although an explosion had occurred in the Unit 1 reactor building, the primary containment remained intact and there would be no major escape of radioactive material. "Please remain calm," Edano told viewers. It was advice the government repeated often.

Absent from this briefing was Koichiro Nakamura, deputy director general at NISA, who had been handling regular updates for the media. Nakamura's candor at press conferences earlier that day apparently had gotten him in hot water with Kan's office. In a morning session with reporters, Nakamura, who had a degree in nuclear engineering, was asked if the fuel had started to melt. He replied: "We cannot deny the possibility." Four hours later, at another briefing, Nakamura said: "It looks like a core meltdown is occurring." A short time later, NISA was notified that the prime minister's office had to clear all statements about the situation, and that NISA also needed to obtain permission from Kan's office before holding further briefings. Nakamura asked to be replaced.

When reporters gathered for another briefing shortly after 9:00 p.m., a new NISA spokesman was in charge. Asked about media accounts of a melt-down, the spokesman backpedaled, saying: "The condition of the core has not been clearly identified yet." The word *meltdown* was not mentioned. The next morning, another NISA briefer was asked about the likelihood of a meltdown. The answer: "The likelihood cannot be denied because such a material [cesium] has already been detected and we must keep that in mind."

In short, if reporters were looking for definitive answers from the government's nuclear safety officials, NISA wasn't providing them. In its steadfast effort to avoid the word *meltdown*, NISA ultimately came up with its own terminology: *fuel pellet melt*. (TEPCO also eschewed *meltdown*, preferring *core damage*.) Although NISA's classification made technical sense, the agency had little basis to avoid concluding that a full-scale meltdown had occurred. In the minds of many listeners, its careful parsing of words was a distinction without a difference.

During the afternoon of March 12, after the explosion in Unit 1, Kan was meeting in his office with Madarame and several others when the discussion turned to the question of injecting seawater into the Unit 1 reactor as a last-ditch attempt to cool the damaged core.

TEPCO officials, realizing that the injection of seawater offered perhaps the last hope of cooling the reactor, had already given the go-ahead. That information was passed along to members of the Emergency Operations Team working in the basement of the prime minister's office building. But the news never made it upstairs to Kan's office.

The decision to use seawater is the death knell for a reactor, and in this case it would send an unequivocal signal that Unit 1 had now been written off by its owners as a piece of radioactive scrap. Seawater is highly corrosive to the internal parts of a nuclear reactor; even the freshwater used in a reactor is filtered to protect components from impurities. The Pacific might cool the core, but it would finish the job of destroying the reactor.

Kan asked Madarame about the possibility that the seawater could trigger recriticality—an uncontrolled chain reaction—in the damaged fuel.[2] Madarame reassured the prime minister that this was unlikely. Participating in the conversation was TEPCO's liaison, Ichiro Takekuro, who interpreted the discussion to mean that the government might not approve of injecting seawater.

Shortly after 7:00 p.m., Takekuro called plant superintendent Yoshida to tell him about the conversation. Yoshida said that the seawater injection had already begun. Takekuro told him to stop it because the prime minister's office was still weighing that option. An angry Yoshida called TEPCO headquarters and was told he had no choice but to suspend the injection pending a decision by Kan.

Yoshida knew that if he stopped the injection, he might not get the jury-rigged pumping system to work again. And he knew that continuing to get water into the core was the highest priority, regardless of the government's apparent dithering. In a moment of bare-knuckled defiance, Yoshida summoned the person in charge of the injection work and whispered instructions: "I'm going to direct you to stop the seawater injection, but do not stop it." Then, in a loud voice, heard in the plant's emergency response center—and, via video feed, in Tokyo—Yoshida ordered that the pumping be halted. Finally, at 8:20 p.m., Takekuro delivered word that the prime minister had given the okay and the injection could resume. By then, however, Yoshida had a new spate of emergencies.

On Saturday morning, March 12, NRC chairman Gregory Jaczko was en route to an 11:00 a.m. meeting at the White House to discuss the situation in Japan when he received an update. Information was still in short supply at the NRC, with most coming via media outlets. The NRC wasn't alone; the Departments of State, Defense, and Energy, the U.S. Agency for International Development, and especially the U.S. Embassy in Tokyo were also laboring to learn more.

Martin Virgilio, the NRC's deputy executive director for reactor and preparedness programs, was manning the Operations Center at White Flint and had just gotten off the phone with nuclear industry officials. They had provided new information that Virgilio, a level-headed thirty-four-year veteran of the NRC, wanted to pass along to Jaczko by phone.

Virgilio's information had come from the Institute of Nuclear Power Operations (INPO), an industry group created after the Three Mile Island accident to bolster nuclear safety initiatives and enhance the industry's tattered public image. INPO, and its international counterpart, the World Association of Nuclear Operators (WANO), had been working sources to learn as much as possible. (One valuable source for the industry and the NRC was Chicago-based Exelon Corporation, which operates more boiling water reactors than any other U.S. utility.) The call provided the NRC with important details, including information gleaned by industry people directly from their contacts at TEPCO. Not all of the information, however, was accurate.

INPO and WANO representatives told Virgilio that a hydrogen explosion had occurred in the Unit 1 reactor building, confirming the assessment of

NRC staff members who had watched the explosion on television. It was a significant detail. If the explosion was in the turbine building, the hydrogen could easily have come from sources other than damaged fuel in the reactor core and/or spent fuel pool. But if the reactor building exploded, it was a far more likely indicator of extensive fuel damage somewhere.

As the industry representatives and the NRC shared what they knew, they pieced together this picture of the situation at Fukushima Daiichi: after the explosion, radiation levels at the plant boundary had jumped to one hundred millirem (one millisievert) per hour—about ten thousand times background—but then dropped to fifty millirem per hour. The primary containment at Unit 1 was believed to be still intact. Iodine and cesium isotopes had been detected, indicating some melting of fuel. About eight inches (twenty centimeters) of the fuel in Unit 1 had been partially exposed, and maintaining water levels in that reactor vessel was problematic. There was an unconfirmed report that the Unit 1 containment was being filled with borated seawater using fire trucks. Unit 3 appeared to be in cold shutdown. Unit 2's RCIC cooling system seemed to be doing its job of transferring heat from the core to the torus, but there was no electrical power to run the pumps that cooled the torus water and transferred heat to the sea. If heat could not be removed from the torus water, the containment temperature and pressure would continue to rise.

Although trouble seemed to be brewing at Unit 2, the Unit 1 situation topped everyone's list of concerns. And in hindsight, they were not wrong—Unit 1's damage was actually far more extensive at that point than the briefing had indicated. However, the worsening conditions at the other two units, as well as Unit 4, would soon vie for the NRC's attention.

For the NRC, the fragmentary information spilling out of Japan posed a combination of challenges the staff had never encountered during emergency drills; nobody had trained for an event this severe. The experts at White Flint and elsewhere within the agency were largely forced to rely on secondhand information and computer analyses that might not have accurately reflected what was actually happening. At the same time, the NRC, as home to the U.S. government's resident experts in nuclear power, was being counted on to provide guidance to ensure the safety of Americans scattered all over Japan, including 38,000 U.S. troops, 43,000 dependents, and 5,000 civilian defense employees, plus embassy personnel, businesspeople, students, and tourists.

One of Jaczko's main concerns, even at that early stage, was the potential radiological impact of the accident and how far away from the site significant radiation exposure could occur. The NRC's Protective Measures Team had already begun to run computer simulations, using a code called RASCAL (the Radiological Assessment System for Consequence Analysis) to address this question. However, as was the case with SPEEDI, the RASCAL simulations were of limited value without data on how much radiation was actually being released. Virgilio told Jaczko that, based on crude assumptions, including a total failure of the Unit 1 reactor containment and severe core damage, the exposure levels that would trigger evacuation by U.S. standards would be exceeded at fifty miles from Fukushima.[3] Jaczko asked if there could be danger even farther away and Virgilio said the NRC hadn't modeled past that point. He didn't mention to Jaczko that the RASCAL code was only good out to fifty miles.

In addition to monitoring what was happening in Japan, the staff at the Operations Center was beginning to focus on a different kind of crisis: a potential public relations nightmare at home. The NRC oversees twenty-three reactors of the same design as the units in trouble at Fukushima Daiichi. The agency had been dragging its feet for years on addressing a host of difficult issues, including revising seismic safety rules. And now the NRC faced the unpleasant task of having to defend its oversight before a newly skeptical public.

Although early news releases offered general assurances about U.S. reactor safety—"Nuclear power plants are built to withstand environmental hazards, including earthquakes and tsunamis"—the NRC didn't know enough about the Japanese plants to say with certainty that the safety situation was any better in the United States. For instance, the NRC did not know at the time whether Japan had installed hardened vents at its Mark I boiling water reactors back when most U.S. plants did so, at the urging of the NRC, in 1989.[4] And even if NRC officials could legitimately claim U.S. plants were safer, they had to tread carefully or risk embarrassing a major U.S. ally. One NRC staffer called this a "slippery slope."

While the NRC was trying to pry more information from Japan, it also faced pressure to be more forthcoming. Some of the push was coming from its own staff. As Jaczko was heading home after the White House meeting, Virgilio asked his boss if the NRC might step ahead of the other federal agencies, "sanitize our Qs and As [about the accident] and post them on the website." Jaczko said no. "At this point, all of the public communication needs to be coordinated," he said. "[L]et's continue to keep it within the federal family."

At that moment, the federal family was engaging in a little sibling rivalry, with the NRC butting heads with the Department of Energy (DOE), which fields its own team of nuclear experts, many of whom work in the national laboratories around the country. Those experts were "chomping at the bit" to get involved, which didn't sit well at the NRC. "[S]omebody might want to call DOE and tell them to tell their labs to cool it," said the NRC's Brian Sheron on March 12, "because the last thing we want is the labs going off, talking to the press, talking about consequences and all sorts of other stuff."

State officials were beginning to press NRC officials for an assessment of radiation risks to their residents.[5] "I think we need to do more with the states," Virgilio told Jaczko. That's where most requests for information were coming from, he said. So far, anyway. The U.S. news media, already plying the NRC for information about events in Japan, were also looking for a local angle.

Eventually, Jaczko and his team knew, the NRC would have to answer the question: could Fukushima Daiichi happen here? For the NRC, *that* was the ultimate "slippery slope."

"I think . . . it's inevitable that we're going to get those questions and calls for comparisons of U.S. facilities to [those in] Japan," the NRC's Brian Mc-Dermott, joining Virgilio on the call, told Jaczko. Jaczko agreed and told his crew to start working on some talking points.

Later that day, the conversation in the NRC's Operations Center turned to crafting acceptable answers, first for the states and then, ultimately, for the media and the public. Staying on top of the message could be difficult, the NRC team agreed.

"[I]f we're trying to restrict the information, or at least control the information, that we're getting outside, anything that you give a state or a governor, you're not going to be able to control what they do with it, even if you ask them to keep it close," said Dr. Charles L. Miller, an NRC veteran who oversaw state programs. Officials in western states, in particular, wanted to know what the NRC knew, especially the likelihood of a meltdown in Unit 1 and the possibility of radiation reaching the United States. So far the NRC had refused to make any public statements about exactly what was happening at Fukushima.

Together, a handful of staff members drafted the talking points. The gist of the message, they agreed, should be this: The NRC will not speculate. Don't believe everything reported in the media "by the so-called experts." Be cautious. "And avoid answering questions on whether it could happen here," Miller warned. "I mean, I don't think we're ready to do that yet."

As the crisis continued, the NRC exerted considerable discipline, refraining from appearing to second-guess the plant operators or to make pessimistic assessments ahead of the Japanese. The press releases it issued, while seeking to reassure Americans, flatly stated that "the NRC will not comment on hour-to-hour developments at the Japanese reactors." As for speculating,

Bill Borchardt, the NRC's director of operations, had his own advice for his impatient colleagues: "Don't . . . scratch the itch."

Within minutes of getting the official go-ahead from Tokyo to "resume" injecting seawater into Unit 1, Yoshida found the next problem landing in his lap. At 8:36 p.m. on March 12, the gauge monitoring water levels in Unit 3 ran out of battery power. In fact, problems in Unit 3 had been quietly multiplying. They, too, involved cooling the core.

Aware early on that battery life was an issue, the Unit 3 shift team had cut off all nonessential power loads, fearful that if the RCIC stopped for some reason there might not be adequate power to restart it. With battery power alone, operators were able to keep the cooling system running by continuously monitoring the water level and controlling the flow rate.

Because Unit 3 appeared to be stable with the RCIC limping along, Yoshida and his workers had focused on restoring power to Units 1 and 2. But at 11:36 a.m. on March 12, the RCIC in Unit 3 stopped and could not be restarted. Now Unit 3 needed water and needed it quickly. All available fire engines were being used to pump water into Unit 1. Obtaining another off-site truck was not immediately possible because roads were treacherous.

Fortunately, Unit 3 had an emergency cooling system that was still functional: the "high-pressure coolant injection" system or HPCI (pronounced *hip-sea*). When water levels dropped in the core, the HPCI automatically kicked in about an hour later, also drawing on battery power. Under normal circumstances, the RCIC was used to deliver makeup water to replace coolant boiled away by decay heat. The HPCI system, with about ten times the flow rate, is designed for much larger water losses, such as could occur from a ruptured pipe or stuck-open relief valve. But these weren't normal circumstances, and the operators worked to control HPCI flow as they had been doing with the now dormant RCIC.

The water level rose and pressure inside the reactor decreased. Yoshida ordered preparations to vent the containment, knowing it was only a matter of time before that became a necessity. He hoped to vent while radiation levels in the reactor building were low.

The last reading of the Unit 3 water gauge, before the batteries died, indicated that the fuel might be covered by a mere sixteen inches of water. Supervisors doled out thirteen two-volt batteries from the stockpile of fifty batteries that had been delivered overnight from another power station, and the gauge was reactivated. Stabilizing Unit 3 became the priority now that Unit 1 was being cooled by seawater injection.

However, the HPCI pressure started to drop and the shift team at Unit 3 began to worry that it could not depend on the HPCI much longer. The team decided to switch to external injection, using a diesel-driven fire pump, which team members thought would be more reliable. But to do this, they would first have to depressurize the reactor vessel by remotely opening a safety relief valve (SRV). (The reactor vessel's pressure had to stay low enough for the low-pressure fire pump to be able to inject water.) The shift team thought opening the SRV was possible because the valve's indicator light was on in the control room, signaling that the SRV could still be operated from the control panel. Yoshida was notified, but he didn't think dealing with the issue was a priority in light of other, more pressing crises.

Shortly before 3:00 a.m. on March 13, the Unit 3 operators decided to shut down the HPCI. However, when they tried to open the SRV, the switch didn't seem to work and the SRV indicator didn't change from green to red to show it was open. The workers attempted to use the fire pump anyway and persuaded themselves from a sound they interpreted as flowing water that it was working. None of this was reported to Yoshida until nearly 4:00 a.m. If he had known in advance of the plan, Yoshida probably would not have approved it: he knew the fire pump would not have worked well even if the SRV had been successfully opened.

After Yoshida discovered what had happened, he decided that the team should try to depressurize the reactor vessel manually, using scavenged car batteries to power the valve controls, and then inject water using a higher-pressure fire engine that had become available from elsewhere on the plant site when a road was cleared. But the reactor had already been without cooling for an hour, and it had entered a critical phase. After a final attempt to restart the HPCI and RCIC failed around 6:00 a.m., Yoshida had no choice but to report to Tokyo that operators had lost the ability to provide emergency cooling to Unit 3.

Although there is some uncertainty about the timing, by approximately 9:00 a.m. on the morning of March 13, the fuel rods became uncovered, triggering the same sequence of events that had occurred at Unit 1 some thirty-six hours earlier: overheating of the fuel, oxidation of the fuel cladding, hydrogen formation, and release of fission products. The Unit 3 core was melting down, and the vessel pressure was rising fast. As though in a nuclear version of *Groundhog Day*, workers once again faced the urgent need to vent the containment to reduce pressure so they could inject makeup water to try to halt damage to the reactor core.

Racing against time, the operators attempted to open the valves in the piping between the containment and an exhaust stack in order to discharge gas, but they couldn't outpace the increase in pressure within the intensely hot reactor. Another reactor now appeared doomed.

Three hours passed before venting got under way and the containment pressure dropped. Because core damage had already begun, high levels of radioactive materials were vented and radiation levels at the plant boundary spiked. But soon afterward, a safety relief valve opened, perhaps automatically, and depressurized the reactor, allowing injection of a newly uncovered supply of freshwater in fire cisterns to the core. Freshwater supplies were running low, however, and Yoshida ordered workers to prepare to switch to seawater. Just after noon, the freshwater reservoir had been depleted and workers scrambled to hook up the seawater line. Seawater began flowing in an hour. But once again it was too little, too late; measurements indicated that the core remained exposed, a sign that not enough water was getting into the reactor vessel to re-cover the fuel. Radiation levels were continuing to rise, hitting 1.2 rem (twelve millisieverts) per hour on the Unit 3 side of the shared control room.

Across Japan, concern was growing about this disaster that seemed to worsen by the hour. Late Sunday evening, TEPCO officials held a news conference. President Shimizu, dressed in a blue company uniform, apologized to the Japanese people and said that the tsunami had "exceeded our expectations." With that, he dropped out of public view for about two weeks, prompting much speculation in the Japanese and foreign media. (TEPCO officials later explained that Shimizu had fallen ill because of overwork.) Shimizu apparently remained involved in decision making, however. He and Prime Minister Kan would soon tangle over a highly publicized misunderstanding.

TEPCO corporate management likely wasn't winning fans inside the frenzied command center at Fukushima Daiichi, either. Tapes of conversations between Tokyo and the center reveal a certain disconnect.

Earlier Sunday, headquarters had ordered one thousand spare car batteries, but delivery to the plant was held up for hours by delays in obtaining the government permits necessary to use the expressways. By Sunday night, with the shipment still en route and batteries in employees' cars and trucks already appropriated, the need for the additional batteries had grown critical. About 7:15 p.m., an announcement blared over the public address system at Fukushima Daiichi: "We are going out to buy some batteries, but we are short of cash. If anyone could lend us money, we would really appreciate it."

And despite his responsibilities managing the deepening crisis, Yoshida's

bosses expected him to keep up with the paperwork. The previous night
TEPCO headquarters had called the exhausted superintendent asking him
to submit his night shift schedules. Yoshida, who at that moment was struggling to get vital water into the Unit 1 reactor, responded: "I'm a bit too beat
to prepare them now." Headquarters persisted and Yoshida eventually replied,
"[N]one of our workers can go home anyway. It's just a question of whether
they are awake or asleep." Whether that satisfied the timekeepers at headquarters is unknown; by the time Yoshida responded to them, they had all gone
home for the night.

Back in the United States, it was early Sunday morning when the NRC's
emergency operations center received a call from Admiral Kirkland Donald,
director of naval reactors at the National Nuclear Security Administration,
a semiautonomous division of the DOE. Donald wanted to notify the NRC
and the DOE, which joined the call, that the U.S.S. *Ronald Reagan*, a nuclear-
powered aircraft carrier off the coast of Japan, had "picked up some activity
out at sea that we think you need to be aware of and probably need to be
addressing with the Japanese government." The "activity" was radiation.

The *Ronald Reagan* and six other navy vessels from the Seventh Fleet had
been en route to South Korea to participate in joint naval activities there
when they were diverted to Japan after the earthquake to help in relief efforts. The ships had taken up position about one hundred nautical miles off
Fukushima Daiichi, with the officers thinking they would be outside any radiation plume. Then radiation detectors in the *Ronald Reagan*'s engine room
picked up readings two and a half times normal.

Radiation was also detected on crew members and helicopters that had
landed on a Japanese ship fifty miles offshore from the plant. Although the
levels were low, the admiral was worried. The measurements were higher
than the navy had expected, based on the scant information available. The
prospect that U.S. personnel might be at risk became real. (There were more
than 3,400 sailors and aviators on the *Ronald Reagan* alone.) The crew and
equipment were decontaminated and the navy ships headed for safer water
about 130 nautical miles offshore. En route, the *Ronald Reagan* continued to
detect contamination.

For the navy vessels, getting out of harm's way was a matter of weighing
anchor and moving farther out to sea. It would not be as simple for the tens
of thousands of Americans living in Japan. They no doubt were beginning to
wonder whether they were safe and where they could go if they weren't.

Approximately 160,000 U.S. citizens were living in Japan in March 2011.

Responsibility for their well-being rested with the U.S. Embassy in Tokyo. U.S. ambassador John V. Roos had put out a call to Washington for guidance almost immediately after the earthquake and tsunami; soon the embassy was also caught up in responding to a nuclear accident. The Japanese were not providing information and relations were growing strained.

The NRC scrambled to get one of its experts aboard a military flight about to depart for Japan late on the evening of March 11. It was such a hasty departure that Anthony Ulses, the chief of the NRC's reactor systems branch, had no time to pack. By March 14, a total of eleven NRC experts had been dispatched to Japan.

It was that same day, March 14, that the Japanese relented and accepted the U.S. offer of assistance, and as a result more systematic data collection began. The National Nuclear Security Administration sent thirty-three people and eight tons of equipment that are part of its Aerial Monitoring System (AMS). The AMS uses specially equipped aircraft to identify and measure radiation in an emergency. Eventually, a U.S. Global Hawk drone went aloft over the reactors to begin gathering data and images that were shared with the Japanese government.

On the morning of March 14, radiation levels at monitoring posts around Fukushima Daiichi were high, and the drywell pressure in Unit 3 was again starting to rise. Workers had successfully vented the containment three times the day before—or at least thought they had—and now needed to do so again. Although they managed to get power and compressed air to the necessary valves by using a portable generator, the drywell pressure continued to increase. The crew was aware that there had been a gap of more than five hours between the time it shut down the HPCI and the time it managed to set up the fire engine to inject water into the core: more than enough time for significant core damage to occur.

Even worse, although water was being pumped into the reactor, the level inside did not appear to be rising. No one had known for quite some time just how much water had been getting into the reactor vessel, because of the intermittent operating histories of RCIC, HPCI, and the fire engine. Workers feared that Unit 3 might now be following a trajectory similar to that of Unit 1. Meanwhile, another crisis was about to emerge in an unexpected place.

While plant personnel focused on cooling the three reactors that had been operating when the earthquake struck, trouble was slowly brewing in one of those that had been shut down on March 11. In the spent fuel pool of Unit 4, the water temperature was steadily rising, inching toward the boiling point. At the time of the earthquake, Unit 4 had been out of service for maintenance, and the entire reactor core was sitting in the spent fuel pool, high above the reactor vessel. Consequently, the fuel pool was loaded with more than 1,500 fuel assemblies. More than one-third of the spent fuel assemblies had been removed from the core just three months earlier, meaning they were still relatively "hot" in terms of both temperature and radiation levels. Of the seven spent fuel pools at Fukushima Daiichi—one for each reactor plus a common pool for extra storage—the Unit 4 pool had the highest heat load by far.

The spent fuel pools are big tanks of water, forty-five feet (fourteen meters) deep, with walls and floor made of reinforced concrete and lined with steel. Spent fuel assemblies are stored in racks in the lower third of the pools. The large volume of water above the racks provides radiation shielding. Without the water to provide shielding, the radiation level even at the rim of the pool would be quickly lethal to workers. Because spent fuel generates heat and high levels of radioactivity for many decades after discharge from a reactor, active cooling systems are critical. During normal operation, pumps circulate water to maintain a pool temperature of about 100°F (40°C). Once cooling is interrupted, the water in the pool will begin to heat up, and if it should reach the boiling point losses can mount rapidly.

The spent fuel pools at Units 1 through 4 had been without power to circulate cooling water for almost three days. Yet until now operators had paid little attention to them. For decades, engineers believed that spent fuel pools posed much less safety risk than operating reactors and hence required far less protection. Compared to fuel assemblies in an operating reactor core, most assemblies stored in spent fuel pools are cooler and less radioactive, because they have been out of a reactor for years and their shorter-lived fission products have decayed away. If a spent fuel pool were to lose cooling, operators typically would have days to respond before the massive volume of water in the pool could heat up enough to boil and eventually uncover the fuel. More severe scenarios—such as a massive earthquake that might damage the integrity of the pool structure itself—were regarded as too improbable to worry about.

This lackadaisical attitude informed the way spent fuel pools were designed and built. Regulators in the United States and elsewhere who reviewed boiling water reactor designs did not require that the spent fuel cooling systems be as robust as those for the reactor core. Nor did they require that the pools be surrounded by leak-tight, pressure-resistant containment structures such as those mandated for the cores. And, in approving GE's Mark I and II reactor designs, the regulators saw no problem in having the spent fuel pool perched on the fifth floor of the reactor building and topped only by a steel frame structure, despite the obvious dangers such an arrangement might pose in a serious accident.[6]

The safety of these pool designs was predicated on the ability of operators to restore cooling within a few days. Apparently little thought was given to the question of what would happen if they couldn't respond promptly—say, if they were occupied with a prolonged loss of power and three reactor meltdowns simultaneously. Neither did the designers or regulators worry much about what would happen if the operators didn't have as much time to act as everyone assumed—for instance, if a large earthquake caused the pool to lose water by sloshing it over the sides or cracking the liner and causing it to leak.

What could happen is this: if the water level did decrease and uncover the fuel rods, they could overheat and melt, much like those inside the reactor—but with one critical difference. The spent fuel is not enclosed by the robust primary containment.

Even though the amount of heat in a spent fuel pool is far lower than that in a reactor, the zirconium alloy cladding encasing the hottest rods can catch fire if it reaches a temperature of 800°–900°C, generating its own heat source. This in turn could rapidly damage the fuel pellets themselves and allow the release of gaseous isotopes, like the long-lived cesium-137. Under certain circumstances, such a fire could spread to other assemblies in the pool. And because typical spent fuel pools hold more fuel than a single reactor core, a fire that involved the entire pool could release more cesium-137 than a reactor meltdown. (Iodine-131, another major isotope of concern in a reactor accident, poses much less risk in a spent fuel pool. It would not be present in significant amounts unless the fuel in the pool was less than three months old; after that time, the iodine-131 would largely have decayed.) And finally, reaction of the zirconium with steam would produce explosive hydrogen gas, just as it did in the damaged reactor cores at Fukushima Daiichi.

At that point, the only barrier between these isotopes and the environment would be the leaky, flimsy top of the reactor building. Such a structure had already proven to be no match for the hydrogen explosion that occurred at Unit 1. When that happened, radiation levels did not increase very much because the primary containment inside the building remained intact. But a hydrogen explosion resulting from a spent fuel fire at Unit 4's spent fuel pool, occurring in a largely uncontained environment, would release a catastrophic amount of radiation.

By the time operators began to focus on ways to cope with the rising temperature in the Unit 4 pool, the radiation level there was already so high that it precluded any human intervention to get water into the pool using temporary pumps. As they began to explore other approaches, the operators knew time was against them.

But on this Monday morning, March 14, the next explosion to hit Fukushima Daiichi came not from Unit 4, but from Unit 3. At 11:00 a.m., a huge blast blew out not just the upper sections of the reactor building, as had happened at Unit 1, but also large sections of the walls, injuring a number of people nearby. Debris fell into the Unit 3 spent fuel pool and onto the ground, where it damaged the fire engine and hoses positioned to inject seawater into the reactor. Once again, hours of painstaking work were undone in an instant.

Smoke pours from Unit 3 following a blast at 11 a.m. on March 14. Falling debris injured workers on the ground and damaged firefighting equipment that was being readied to inject vital cooling water into the reactor. *Tokyo Electric Power Company*

By early afternoon, radiation levels above thirty rem (three hundred millisieverts) per hour were detected just north of Unit 3, raising concerns that the core had already undergone significant damage and that the explosion had dispersed radioactive material around the site. The rising radiation also rendered part of the Units 3 and 4 control room unsafe, further hampering operators' ability to gauge the condition of the reactor.

Compared to the seriously damaged Units 1 and 3, Unit 2 had seemed like a success story. In contrast to Unit 3, the RCIC at Unit 2 had apparently continued to operate on its own for nearly three days, keeping the water level from dropping and exposing the core. This was remarkable, because ordinarily the RCIC should not have been able to operate for that long without battery power. But operators knew it couldn't keep running indefinitely.

The RCIC transferred heat from the reactor vessel to the torus, but that heat remained within the containment, which itself was already overheated. Without any means of external cooling, pressure and temperature in the containment continued to increase. The RCIC had not been designed to function under such punishing conditions. And at around 1:30 p.m. on March 14, it looked like it had finally quit. The Unit 2 water level started to drop. Now it was clear that, without prompt action, Unit 2 would suffer the same fate as Units 1 and 3.

For exhausted shift operators, some of whom had been on duty for three days without sleep, the crises seemed to have no end. The operators needed to inject water into the Unit 2 reactor vessel as soon as possible to keep the core from being exposed. But just as at Units 1 and 3, they first needed to vent the containment. Otherwise, they would not be able to depressurize the reactor vessel adequately to force water into it with the fire engines. Yoshida had anticipated this eventuality right after the Unit 1 explosion and had ordered workers to prepare the means both to vent the Unit 2 containment and to inject water into it when needed.

But Yoshida had not anticipated the Unit 3 explosion, which disrupted both the Unit 2 containment vent line and the alternate water injection line that workers had established. This setback marked the third strike against the doomed plant. Working amid rubble, aftershocks from the earthquake, and high radiation levels, a crew raced against time to repair the damage to the Unit 2 vent line and to construct a new water injection line. Workers tried to vent the containment at 4:00 p.m. but failed. Meanwhile, the water level inside the reactor vessel kept dropping. By 4:30, the top of the core was exposed.

TEPCO president Shimizu ordered Yoshida to try to inject water into the reactor vessel without waiting for venting to succeed. But to make the attempt, workers needed to open a safety relief valve that would move steam from the reactor vessel into the torus. Efforts to open the valve were complicated by a lack of battery power. (This relief valve is different from the containment vent valves that workers struggled with on Units 1 and 3.)[7]

Once again, a search was mounted for batteries to energize the safety relief valve. These were carried to the control room and connected, but they lacked enough voltage. The water level continued to drop. Finally, at about 6:00 p.m., additional batteries allowed the valve to be opened. But the water was now so low it didn't even register on the gauge, meaning that the core might be completely uncovered. Also worrisome was the fact that, although hydrogen and radioactive gases were flowing into the containment, pressure there was not rising, as it had in the other units. The containment possibly had sprung a leak, allowing gases to escape into the reactor building and out through the hole blown in the side of the building when Unit 1 had exploded.

By 7:00 p.m., pressure inside the Unit 2 reactor had dropped far enough for the fire engine pump to force water into the core. But a passing worker discovered that the fire engine, which had been idling while awaiting the injection operation, had run out of fuel. The engine was restarted shortly before 8:00 p.m., but water levels in the core remained too low to measure. Soon pressure inside the reactor rose again, thwarting the fire engine pump. The intense heat from the exposed fuel was quickly vaporizing the water being injected. As steam refilled the reactor vessel, the pressure increased. To enable the pump to inject water again, the operators managed to open a second relief valve. That appeared to do the trick; water inched up to cover part of the fuel. But the progress came at a cost: steam pressure inside the reactor was on the rise too. For the next several hours, the operators played cat and mouse with water level and pressure. The steam pressure periodically spiked above the water pressure of the fire engine hoses, and the operators could not be certain how much water was getting into the reactor vessel.

At the same time, operators were trying without success to vent the Unit 2 containment. According to the gauge readings, by now the containment pressure should have forced the vent open by blowing through a rupture disk, but it had not. The pressure in the drywell suddenly began to rise rapidly, and Yoshida was afraid it would burst. He decided it was necessary to try to vent the drywell directly—an action that was considered a last resort because it would allow radioactive gases to escape into the environment without first being filtered through the water in the torus. However, for better or worse, operators were not able to vent the drywell either.

The operators continued to struggle through the early hours of March 15 to find a way to vent the containment. At 6:00 a.m., just as a shift change in the Units 3 and 4 control room was taking place, a noise that sounded like an explosion was heard in the area around the torus beneath the Unit 2 reactor vessel. Pressure in the torus dropped, suggesting that the explosion had damaged the torus and allowed its contents to escape into the reactor building. If the containment had actually been breached, vast amounts of radiation could soon spill out into the environment.

Then, around the same time, a hydrogen explosion ripped through the Unit 4 reactor building, collapsing the top two stories of the five-story structure. The incoming and outgoing control room crews fled to the safety of the Seismic Isolation Building, retracing their route on foot this time because the concrete and rubble covering the ground near the reactor made car travel impossible. In their cumbersome radiation suits and breathing equipment, it took them nearly two hours to make their way to the building and alert Yoshida of the damage to Unit 4.

At about 6:20 a.m. on March 15, a huge explosion rips apart the top two stories of Unit 4, exposing the spent fuel pool. Until then, Unit 4 had seemed to pose the least threat in the first days of the accident. *Tokyo Electric Power Company*

The workers at Fukushima had experienced many surprises in the past few days, but the explosion at Unit 4 must have come as a particular shock. Although they were aware that the temperature was rising in the spent fuel pool, they believed that Unit 4 was the least of their problems. Now it appeared that the pool had heated up much faster than they had expected, so rapidly in fact that the spent fuel might have become uncovered, sustained damage, and generated explosive hydrogen. But even more worrisome was the realization that the other spent fuel pools at the site might also be in jeopardy—a particular concern given that the pools at Units 1 and 3 were now directly exposed to the environment. In an instant, the spent fuel pools went from being a low priority to an immediate threat.

In response to the deteriorating situation, at 11:00 a.m. Prime Minister Kan ordered residents between about twelve and eighteen miles (twenty to thirty kilometers) of the reactor to remain indoors. Those who hadn't already fled now became housebound, relying on their small frame homes to protect them from whatever might happen next, and not knowing when they might be allowed out again. No one had ever warned them to prepare for a situation like this.

With the catastrophic events at Fukushima Daiichi showing no sign of slowing, the radiation exposure levels of the workers grappling with the disaster had begun to worry government and utility officials in Tokyo. The government came up with a solution: increase the allowable limits of exposure.

On March 14, the government agreed to raise the maximum allowable radiation dose for workers in an emergency from ten rem (100 millisieverts) per year to twenty-five rem (250 millisieverts) per year. The decision, announced the following day, was based, in part, on fear that soon everyone at Fukushima Daiichi might exceed the existing exposure limit and thus be required to leave. Already workers had to perform tasks in short bursts to stay within cumulative dose limits.[8]

That evening, TEPCO president Shimizu was told that, because of the worsening conditions, Yoshida was considering evacuating about 650 nonessential personnel from the plant. Shimizu telephoned the head of NISA before dawn on March 15 to apprise him of the grave conditions at Unit 2 and noted that if things continued to worsen, the company might pull out some personnel. In another instance of bungled communication, Shimizu apparently did not make it clear that essential workers would remain.

Officials in the prime minister's office feared a total abandonment of the plant, in which case the dangers would grow exponentially. At around 4:00 a.m., an angry Kan summoned Shimizu to his office, where the utility chief told him there was no plan to abandon the plant. (The media reported that Kan, irate that he was not informed of events at the plant, shouted at

Shimizu, "What the hell is going on?") The encounter apparently convinced both Kan and Shimizu that the lines of communication—and cooperation—needed to improve quickly. The two men agreed to establish a government-TEPCO integrated response center at the utility headquarters.

About ninety minutes later, Kan visited TEPCO headquarters to make it clear that the government was asserting a stronger role in the accident response. En route, he said later, he considered a worst-case scenario in which evacuations could extend as far as metropolitan Tokyo, population 13 million. (Indeed, mass evacuations might have been necessary, according to a worst-case assessment prepared at Kan's request by the chairman of the Japan Atomic Energy Commission and presented to the prime minister on March 25.)

The TEPCO showdown coincided with the arrival of the latest bad news from Fukushima Daiichi: the possible containment rupture at Unit 2, followed by the blast in the Unit 4 spent fuel pool.

When the Japanese finally did ask for help from the United States on March 14, they were looking for heavy-duty pumping equipment that could deliver seawater to pits from which it could then be injected into the reactor cores via fire engines. U.S. military forces in Japan had firefighting equipment that would fit the bill, and discussions already were under way with TEPCO.

Over the course of that day, NRC staffers began arriving in Tokyo, including Tony Ulses and Jim Trapp, both experts in boiling water reactors. Finally, the NRC's emergency operations center at White Flint had eyes and ears on the ground. The NRC team joined experts arriving from other government agencies, most of them based at the U.S. Embassy.

Later on Monday, Jack Grobe, the NRC's deputy director for engineering, was manning the White Flint Operations Center when Chairman Jaczko telephoned for an update. Grobe had worrisome news: the prevailing winds in Japan, which had been carrying radiation out to sea, were about to shift to the southwest—toward Tokyo. With three reactors and the Unit 4 spent fuel pool in trouble, "that changes the dynamic of the protective measures aspects of this," Grobe told his boss. Vast numbers of people could now be downwind. As a result, the Operations Center had summoned additional staff experts on radiation dose assessments and meteorology. But so far, Grobe told Jaczko, Japan's 12.4-mile (twenty-kilometer) evacuation zone around Fukushima Daiichi was "consistent with what we would recommend."

Marty Virgilio then called Grobe to tell him U.S. nuclear industry officials now speculated that there was a crack in the torus at Unit 2. If true, this could dramatically change the team's assessment of the situation. The torus was part of the primary containment. Up to that time, as bad as things had gotten, the NRC was not aware of evidence that the primary containments at any of the reactors had experienced a major rupture. Such a breach could release dozens of times as much radioactive material as the minor leakage that had previously occurred. Given the projected wind shift inland, a Unit 2 containment failure could have catastrophic consequences for a large segment of the Japanese population.

"I mean, this is beginning to feel like an emergency drill where everything goes wrong and you can't, you know, you can't imagine how these things, all of them, can go wrong," said Grobe. No drill had ever come close to this.

But the bad news hadn't stopped arriving. Jim Trapp called to say that an admiral at the U.S. naval base at Yokosuka, south of Tokyo and 188 miles from Fukushima Daiichi, was reporting radiation measurements of 1.5 millirem (0.015 millisievert) per hour, apparently because of the wind shift. The NRC team was astonished that such a high dose rate would be detected at such a great distance from the plant.

Trapp added one more ominous bit of information from Japanese officials about the blast inside Unit 2: "[T]hey do believe they breached the primary containment." Tony Ulses, also on the phone, raised an even more frightening prospect: the loud noise heard in Unit 2 "was probably when the core went X-up"—meaning when it had melted through the bottom of the reactor vessel. "Landing in the water under the vessel, it would have caused a little steam explosion," said Ulses.

"Believe it or not, Jack, we're telling you the good news," Trapp told Grobe. Then he launched another dire litany, based on information gleaned from NISA officials in Tokyo. No one was certain of the water level in the spent fuel pool at Unit 3; as a result of the explosion, the reactor building "really collapsed into the fuel pool." Radiation of ten rem (one hundred millisieverts) per hour was being measured near Unit 4, which could prove lethal after a day or two of exposure. (For context, workers at U.S. nuclear reactors are limited to five rem per year and generally receive far less than that.)

In an attempt to find some good news to pass along to his exhausted colleagues in Tokyo, Grobe told Trapp and Ulses that additional NRC staff members were en route. But that prompted a word of warning from Trapp: "[W]e've got to think about . . . whether we want them to come."

"I mean," added Ulses," we're getting to the point where this is just more bodies to have to get back out of here possibly."

4

MARCH 15 THROUGH 18, 2011:
"IT'S GOING TO GET WORSE . . ."

Jim Trapp and Tony Ulses were back on the phone with their colleagues at White Flint before dawn East Coast time on March 15. The original plan for the pair to spell each other off until reinforcements arrived had fallen apart; there was just too much going on. Both were exhausted; it had been a roller-coaster day.

At first, it had seemed that the flood of bad news from Fukushima Daiichi was slowing. The latest pressure readings indicated that the core of the Unit 2 reactor had not breached the reactor vessel, and operators were able to reestablish water injection. Other pressure readings contradicted the assumption that the mysterious noise heard at Unit 2 had been a rupture of the torus, although the data were inconsistent, suggesting an instrument failure. Units 1 and 3 were stable, with seawater injections proceeding smoothly.

The details even elicited a bit of optimism from Brian McDermott a short time later when he briefed his boss, Gregory Jaczko. "[L]ast night we thought we had a big problem," he told Jaczko, referring to Unit 2. "And this morning, it, it suggests we have less of a problem."

That assessment was short-lived.

Overnight, concern about the status of Unit 2 had led the NRC's Protective Measures Team to run computer simulations for that unit based on worst-case assumptions: a fully molten core and a ruptured containment. The simulations projected levels of radiation exposure based on the prevailing winds blowing toward Tokyo. Readings picked up at the Yokosuka naval base and elsewhere seemed to suggest that the release was worse than everyone had thought.

Because the NRC's RASCAL program can only estimate radiation doses within fifty miles of a release site, it was of little use in interpreting the data collected at Yokosuka, about 190 miles from the plant. Even so, the findings were passed along to the State Department and the U.S. Embassy in Japan. The calculations indicated that, if winds continued to blow steadily in one direction, then evacuation would be warranted for everyone up to fifty miles downwind of the plant. This was based on the U.S. Environmental Protection Agency's (EPA) protective action guides, or PAGs. (According to the PAGs, members of the public should be evacuated from any area where they could receive more than one rem, or ten millisieverts, of radiation exposure in a four-day period.) Based on the RASCAL estimate, the NRC was recommending to State that U.S. citizens evacuate if they were within fifty miles of the plant, four times the distance the Japanese were recommending. Approximately three hundred Americans resided inside this fifty-mile zone, along with about 2 million Japanese.

From a health standpoint, it seemed a prudent call. Politically, however, a recommendation so at odds with what Japan was telling its citizens was sure to upset Tokyo. The State Department was in an awkward position diplomatically, hoping the NRC might find a way to support the twelve-mile evacuation with a press release—even suggesting the language it wanted the NRC to use, reaffirming the Japanese evacuation recommendations. The NRC's experts stuck by their guns: ask Japan to tell us why we're wrong, they countered.

For the NRC, a fifty-mile evacuation advisory for Japan could also have political ramifications at home. The commission currently requires that emergency evacuation plans be developed only for the area within ten miles of a nuclear reactor. Safely evacuating an area twenty-five times as large would be difficult, if not impossible, at many reactor sites without detailed advance preparation—and plenty of time.

While debate continued about the status of Fukushima Daiichi Unit 2 and its implications, Trapp and Ulses had more bad news to report: the crisis at the Unit 4 spent fuel pool. It appeared, they said, that the situation could be just as dangerous—and baffling—as that in Unit 2, if not more so.

The hydrogen explosion at Unit 4 had occurred almost simultaneously with the suspected explosion in Unit 2 (about 6:00 a.m. March 15, Japan time), but the NRC heard nothing about a problem at Unit 4 until a couple of hours after the report of a possible containment breach at Unit 2. (Masao Yoshida at the plant also had been late in learning. He had not known of the explosion for almost two hours—until his workers made it back to the

Seismic Isolation Building.) The Unit 4 reactor, out of service with its fuel in the pool, had not even been on anyone's radar. In fact, only a few hours before the Unit 4 explosion, Marty Virgilio had reported in a status briefing that there were no concerns about any of the spent fuel pools, although given the extended loss of power, the team needed to "keep an eye on" them. Even after information about the Unit 4 crisis began trickling in, Unit 2 remained at the top of the priority list—but not for long.

As the NRC staffers pieced together data from a variety of sources, a disturbing picture began to emerge that challenged the team's complacency toward the spent fuel pools. The massive blast that had blown off the Unit 4 reactor building's roof and damaged walls there looked like yet another hydrogen explosion. But where had the hydrogen come from? The most obvious source would be a chemical reaction between steam and overheating fuel rods in the spent fuel pool. But that would mean that the water level had dropped to expose the fuel much more quickly than anyone anticipated.

The spent fuel pool was designed to hold 1,300 to 1,400 tons of water, approximately half the volume of an Olympic-sized swimming pool. If the pool lost cooling, it would have taken a couple of days for the water to reach the boiling point. When the water began boiling away, the pool would lose around one hundred tons of it per day. About one thousand tons of water would have to boil off before the fuel would be exposed. By a back-of-the-envelope calculation, it should take well over a week to get to that point. At the time of the explosion, the cooling pumps for the Unit 4 pool had lacked power for less than four days. If the spent fuel had become uncovered in such a short time, then more than eight hundred tons of water must have been lost in some way other than boiling. This could have been a result of shaking during the earthquake. (One worker who was on the Unit 4 refueling floor when the quake struck later reported seeing waves in the pool and being drenched as water sloshed out.) There was another possibility: a leak or crack in the pool itself, caused by the earthquake or the explosion in Unit 3.

No one could verify what was actually going on in Unit 4, or in any of the other spent fuel pools for that matter, because the gauges that measured the water level and temperature in the pools were useless without electrical power.

The worst case seemed to be confirmed when the NRC started to receive reports that a few hours after the blast a fire was observed burning in the

Unit 4 reactor building. Initial fears were that this showed the zirconium cladding on the fuel had indeed ignited, making it possible that the fuel itself was melting and releasing a massive amount of radioactive cesium. With the roof and walls of the Unit 4 reactor building blown apart, the radiation would go straight into the atmosphere. No one could get close enough to investigate; however, the fire burned itself out shortly before noon. Its source was subsequently identified by NISA as lubricating oil used in a generator on a level below the fuel pool. The quake might have damaged piping or a storage tank, and a spark apparently set the oil ablaze. If the zirconium cladding had been burning, the fire would have lasted for days.

Radiation measurements between the Units 3 and 4 reactor buildings had soared to forty rem per hour, making even brief forays near the buildings extremely dangerous. They also seemed to indicate that, even if there had not been a zirconium fire in the Unit 4 spent fuel pool, something bad had happened there: perhaps pieces of spent fuel had been dispersed by the explosion, or there was no water left in the pool to shield the fuel.

Meanwhile, in the hours after the mysterious event at Unit 2 and the explosion at Unit 4, radiation at the plant gate had spiked to its highest level since the accident began, and an ominous cloud of "white smoke" was seen drifting from Unit 2. The prevailing winds continued blowing inland, toward populated areas. News that the wind direction was rotating clockwise, away from Tokyo but toward smaller cities and agricultural areas northwest of the plant, was small consolation. Since it was no longer clear that there had been a containment rupture at Unit 2, reports of white smoke notwithstanding, it was looking more plausible that Unit 4 was the source of the increased emissions.

Now the NRC had a new challenge: trying to understand what was going on with the Unit 4 spent fuel pool. The question of whether the pool still held any water had serious implications for estimating the amount of radiation that potentially could be released. This would soon become a thorny issue for the NRC team.

There was no doubt at White Flint that water had to be added to the Unit 4 pool immediately. That meant plant operators would have to figure out a way to deliver tons of water per hour into a spent fuel pool five stories off the ground, filled with rubble, and emitting levels of radiation that could be lethal within minutes. In the NRC's view, getting water into the spent fuel pools should become the number one priority at Fukushima Daiichi. However, the Japanese did not yet see it that way.

WHERE FUKUSHIMA DAIICHI GOT THINGS RIGHT

Fukushima Daiichi focused world attention on spent fuel pools and their associated risks. Fukushima also demonstrated a safer means of managing the spent fuel risk: dry storage.

There were 408 spent fuel assemblies in dry storage at Fukushima. Although jostled by the earthquake and submerged temporarily by the tsunami, these assemblies survived without the need for helicopters dropping water from above or fire trucks spraying water from below.

Damage inside the dry cask storage building at Fukushima Daiichi was extensive. Although the casks (one is visible, center) were jostled and submerged by the tsunami, they remained unharmed. *Tokyo Electric Power Company*

Dry storage was first used in the United States in 1986, and it is now practiced at plants across the country. After spent fuel assemblies have stayed in a pool for an initial period, workers transfer them into dry storage on-site. Typically, about fifteen tons of spent fuel are placed inside a sealed metal canister, then placed within a concrete and steel cask (the core at a large reactor holds nearly one hundred tons of fuel). Passive cooling—air flow from the chimney effect—removes the heat produced by the fuel.

Passive cooling does not work for densely packed spent fuel pools. The heat produced by the fuel assemblies—particularly those discharged from the reactor core within the past five years—requires continual cooling of the water in the pools.

A large radiation release from a pool could result in thousands of cancer deaths and hundreds of billions of dollars in decontamination costs and economic damage.

The crowded spent fuel pools at U.S. reactors pose hazards. An accident or terrorist attack could cause a loss of water from a pool or interrupt vital cooling systems. As Fukushima Daiichi demonstrated, it can be challenging to pump water up five levels into a pool for a boiling water reactor. And unless a hydrogen explosion rips apart walls and roofs, it can be difficult to drop water into a pool from above or spray water in from the side.

Transferring more assemblies into dry storage is the better way to manage risks posed by spent fuel. Reducing the spent fuel pool inventory accomplishes four things: It (1) reduces the heat load in the pool, (2) adds more water to the pool for every assembly removed, (3) allows the remaining assemblies to be spread out, and (4) reduces the amount of radioactive material in the pool.

The first two of these give workers more time to intervene in the event of a problem, increasing their chances of success. The third restores margin against inadvertent criticality within a spent fuel pool and provides additional space between assemblies for cooling air or water flow. And the fourth limits the size of the radioactive cloud that can be emitted from a pool. Thus, both the probability and the consequences of a spent fuel pool accident are lowered by moving fuel assemblies into dry storage.

There are lessons to be learned from what went wrong at Fukushima. There are equally important lessons to be learned from what went right.

NRC chairman Gregory Jaczko checked in with the Operations Center at White Flint at around 9:00 p.m. on March 15. Jack Grobe was on duty and gave his boss a rundown. Fresh NRC troops were due to arrive in Japan to provide relief to Jim Trapp and Tony Ulses. (By now, about two hundred people from a variety of federal agencies had arrived at the U.S. Embassy in Tokyo to assist.)

Reliable information was still hard to come by, Grobe told Jaczko, and the NRC was relying on "snippets" of information from NISA and U.S. and Japanese news media. "It's been extremely frustrating," Grobe said, adding that he hoped to glean more from industry sources in the United States. Also, the DOE's Radiological Assistance Program teams were now on the ground and would soon be providing crucial land and aerial monitoring. That, Grobe said, "will be a huge benefit."

Jaczko reminded Grobe that it would take at least two hours to obtain high-level authorization if an expanded evacuation order for U.S. citizens was deemed necessary. "I don't anticipate a need for that," Grobe told his boss. "[T]hings seem to be reasonably stable."

Jaczko was due on Capitol Hill the next day for hearings in the House and Senate. He and his staff had been briefing lawmakers since March 11, but this would be the first opportunity for members of Congress to question the chairman in public. Media coverage was a certainty.

The day wasn't over yet. At about 10:00 p.m., the NRC team watched Japanese television footage of smoke pouring from Unit 3. TEPCO announced that radiation levels were rising. A short time later, U.S. media began reporting that TEPCO was pulling its workers from the plant. Although it was unclear if that meant everyone, this was extremely troubling news. Without sustained intervention, the plant condition could quickly deteriorate.

Jim Trapp and Tony Ulses, who had called in to White Flint, confirmed that radiation levels were high. "Unit 4 is in shambles," Trapp told his colleagues. He handed the phone to one of Ambassador John Roos's assistants, who asked if the NRC headquarters knew more about the status of workers at the plant. Marty Virgilio read aloud a *Washington Post* bulletin: "The skeleton crew remaining at Fukushima Daiichi Nuclear Power Plant is being evacuated because of the risks they face from dangerous radiation levels." (This wasn't quite accurate; it later turned out that the workers had been pulled from the heavily contaminated control rooms but remained on-site, available to return if needed. Nonessential personnel had departed the plant, but they eventually returned.) The line went dead briefly and then Ulses explained the interruption: "We just had an earthquake here."

Inside the Seismic Isolation Building, superintendent Yoshida and his fellow workers often had to rely on reports from the emergency crews making brief forays out onto the plant grounds to know what was taking place. Only then were they able to assess just how bad things had gotten. After one explosion, the men inside could hear debris raining down on the roof of the bunker-like building.

"I really felt we might die," Yoshida told a reporter in a rare interview, nearly a year and a half after the accident.[1]

Working conditions in the Seismic Isolation Building rapidly worsened. Personal dosimeters, used to measure radiation levels, were in short supply. Many had been destroyed by the tsunami or rendered useless when their batteries died. Those still functioning were reserved for workers forced to venture outside. Protective clothing also became scarce. Meals consisted of biscuits and dried food, and water was rationed. Crews slept in the hallways.

To support his people, Yoshida said, he passed out cigarettes in the heavily used smoking room. He encouraged them to record their names on a large whiteboard as a memorial in case they did not survive. In the circumstances, he told the interviewer, "it was clear from the beginning that we couldn't run."

Tapes from the videoconference link between the Seismic Isolation Building and TEPCO headquarters, released by TEPCO more than a year later, reveal a chaotic scene. Tempers flare as Yoshida and his team are besieged with orders and counterorders from TEPCO headquarters, which was also getting instructions from the prime minister's office.

Yoshida would later tell government investigators he believed TEPCO's chain of command during the accident was "disastrous." When he was ordered on March 12 to suspend injecting seawater—the order he ignored—Yoshida said he thought: "What the hell is the office talking about?"

By March 18, his patience had all but run out. When officials at headquarters directed him to send crews out to conduct another check, Yoshida snapped: "My people have been working day and night for eight straight days. . . . I cannot make them be exposed to even more radiation."

In the wee hours of March 16, East Coast time, Charles "Chuck" Casto called White Flint from Tokyo, where he had arrived a short time before. Casto, a deputy regional administrator in the NRC's Atlanta office, had been chosen to head the NRC team in Japan.[2] (He would spend almost a year there.) Along with eight other NRC staff members, Casto joined Trapp and Ulses in the U.S. Embassy. The NRC crew set up shop in a large conference room.

Left to right: Charles Casto, Bill Borchardt, and Marty Virgilio. *U.S. Nuclear Regulatory Commission*

The first thing that Dan Dorman, in charge overnight at White Flint, asked Casto for was an update on the Unit 4 spent fuel pool. Casto promised to get what he could. "As you know," he said, "the communications channels are very limited." Even though he had arrived so recently, Casto had picked up the high level of concern pervading the embassy. "[T]hey're worried about the 170,000 [*sic*] Americans over here," he said. "That's their primary goal. They're here to protect them. . . . [T]hat's got to be our focus."

Marty Virgilio, who was on the line in White Flint, concurred. "[T]he top priority is to support the ambassador and the U.S. citizens there. I would say second is support for the Japanese government in the recovery of these reactors. And, then, third is gathering insights and lessons learned for us, so that we can assess the implications for our [reactor] operating fleet and applicants."

Late the previous evening, a RASCAL simulation had estimated radiation doses from a spent fuel fire in a pool that had lost all its water, as well as its roof, and released 50 percent of its radioactive contents to the atmosphere. Just as for the case of Unit 2 with no containment, the calculation indicated evacuation would be needed at least out to fifty miles downwind, RASCAL's limit. So now two possible scenarios would warrant a fifty-mile evacuation, and the NRC still had no way to know whether either was occurring.

But a decision had to be made soon on evacuation advice to Ambassador Roos, and the NRC team needed to judge whether its disaster scenarios were plausible. There would be hell to pay if the United States expanded the evacuation zone and it turned out that the Unit 2 containment remained intact and there was no spent fuel fire at Unit 4.

The lack of solid data about the status of Fukushima Daiichi frustrated everyone. "It was like trying to investigate a homicide and not having access to the crime scene," Casto would later say. "There was just so much misinformation, lack of information, and you didn't know what information to trust." The NRC crew at White Flint agreed to try harder to get details out of the U.S. and international nuclear industry. GE Hitachi, for instance, had its own emergency operations center in Tokyo; if need be, Casto suggested, Virgilio could ask top-level GE officials in the United States for access.

Casto had spent a portion of the fourteen-hour flight to Tokyo buried in a couple of NRC technical publications (known as "NUREGs") relevant to the crisis. For decades, the NRC has funded contractors at U.S. national labs to run computer simulations of possible reactor accidents. Casto was struck by how closely these simulations matched the real-world situation at Fukushima. As he told his colleagues, "of course, that Mark I containment is the worst one of all the containments we have . . . this NUREG tells you that in a station blackout you are going to lose containment. There's no doubt about it."[3]

About ninety minutes later, John Monninger, an engineer with Casto's group, called White Flint to report that the NRC team had been invited to a meeting with TEPCO and Japanese nuclear regulators. TEPCO apparently wanted guidance on dealing with spent fuel pools. Monninger also shared bad news: the pools in Units 1 and 2 were "boiling down." Those in Units 3 and 4 were displaying zirconium-water reactions, meaning that the fuel was exposed. Further, the explosion at the Unit 4 reactor building had "leveled the walls, leveled the structure for the Unit 4 spent fuel pool all the way down to the approximate level of the bottom of the fuel," Monninger said. "So, there's no water in there whatsoever."

TEPCO was talking about dumping sand in the pool, which would reduce the radiation levels but do nothing to cool the overheated fuel rods. The NRC group agreed that a combination of water and boron (to reduce the risk of criticality in the pool) seemed the best answer. The headquarters team members promised Monninger that they would come up with further suggestions before their colleagues in Tokyo left for the meeting in about forty-five minutes.

While the NRC was projecting radiation dose levels under various scenarios, the DOE's Office of Naval Reactors was doing its own assessments, and both agencies were feeding their recommendations to the U.S. ambassador. The two groups weren't seeing eye to eye, it seemed. To the Office of Naval Reactors, the radiation readings detected on naval vessels and at bases far from Fukushima indicated that the event was much worse than the NRC had painted it. While the NRC was estimating the radiation release from damaged fuel in a single reactor, experts at Naval Reactors saw a far greater danger; they believed the NRC was "undershooting" the potential release. As Stephen Trautman, deputy director of Naval Reactors, told the NRC by phone, "[I]f we end up losing one of these plants, there's a good possibility we're going to lose all of these plants." The two sides agreed that the threats from the spent fuel pools also had to be factored in.

For U.S. officials, the evacuation decision was a delicate balancing act. On March 13, Ambassador Roos had put out a press release telling Americans in Japan to "follow the instructions of Japanese civil defense authorities," who were evacuating only a 12.4-mile (twenty-kilometer) zone around Fukushima. Now the ambassador was getting information that suggested the advice might not be sufficient. Yet if the United States were to unilaterally launch a wider evacuation, it could anger an important ally.

Within the embassy itself, fears were high. When Casto made trips to the cafeteria, he was often stopped by staff members, including pregnant employees worried about their unborn children, asking: how bad is it? "There was significant concern," he recalled.

To arrive at a unified recommendation for Ambassador Roos, a high-level phone conference was arranged with Naval Reactors and the NRC teams in Japan and White Flint. During the call Jaczko asked: "[D]o we think this is going to get better or get worse?" Marty Virgilio responded, "It's going to get worse if you think about the spent fuel pools. Right now, Unit 4 doesn't have a spent fuel pool anymore." Water levels couldn't be maintained in the Units 1, 2, and 3 fuel pools, he added.

At worst, Jaczko summarized, the accident could involve three reactors "out of control" and possibly up to six spent fuel pools. The ambassador should base any evacuation decision on that information, he said, and then asked: "Does anybody disagree with that?"

"Chairman," said Admiral Kirkland Donald of Naval Reactors, "I agree with you."

"If this happened in the U.S.," added Bill Borchardt, the NRC's executive director and a veteran of the U.S. nuclear navy, "we would go out to fifty miles. That would be our evacuation recommendation."

It was about 6:30 a.m. Wednesday, March 16. It was time to call the White House.

Jaczko's summary to personnel in the White House Situation Room was succinct. In even terser language, a White House official repeated Jaczko's report to confirm: "Three reactors melting down, six spent fuel ponds go up in flames. And what, then, would be the impact for Tokyo, for example, if the wind kept blowing in that direction?"

"At this point, I think I would still go with the fifty miles right now," said Jaczko, noting the lack of solid information. However, he cautioned that even if the estimates were only "off by a little bit," the impact could be "significantly larger."

Before Jaczko's scheduled 9:30 a.m. testimony on Capitol Hill, he and Borchardt, who was accompanying the chairman, called in to White Flint for updates. Which posed the larger threat now, Jaczko asked his team: Unit 2 or the spent fuel pools? The pools, he was told.

After Jaczko hung up, Borchardt remained on the line to seek more details. Casto, participating from Japan, said that in addition to the threat posed by the Unit 4 spent fuel pool, one had to assume that conditions at the other endangered reactors and spent fuel pools would also deteriorate because the NRC team wasn't seeing any "mitigation" of the ongoing crisis.

Borchardt asked: "I just want to make sure that no one takes anything he [Jaczko] says as implying that, if they resolve the issues with Unit 2, life is going to be a lot better. Right?"

"No," someone else on the call replied.

Discussions at White Flint now focused on how best to deal with the fuel pools. The Japanese were talking about using a helicopter to fly over Unit 4 and dump sand or water into the gaping hole where the roof of the building had been. Casto was growing impatient. "I don't know what they're waiting for."

Radiation levels at sixty-five feet (twenty meters) above the open pool would be "hundreds of thousands of rads," came the reply. Helicopter pilots who flew over Chernobyl to dump sand on the burning reactor there "flew until they died, basically," said one member of the NRC team.

Borchardt called again to "triple-check" the fifty-mile evacuation recommendation. Was it still a go? Charlie Miller said yes, based on current meteorological conditions. But if conditions changed, even a fifty-mile evacuation might not be sufficient. If the wind shifted, Miller added, even Tokyo—150 miles away—could be affected.

A short time later, Monninger called White Flint to report that two government ministry employees had asked for help acquiring emergency equipment to move water from the ocean to the reactor buildings, plus water cannons and four or five trucks with aerial booms capable of spraying water at a height of sixty-five to one hundred feet (twenty to thirty meters) to reach the fuel pools. All of this apparatus would need to be positioned quickly because the radiation level near the reactors was thirty rem (three hundred millisieverts) per hour. Robots with remote cameras and radiation measuring equipment also would be useful, the Japanese said.

"One thing, we went back and forth on [the] Unit 4 [spent fuel pool]," said Monninger, ". . . because we believe the walls were blown out and the water level is at the bottom of the active fuel. That took them back, you know, quite a bit. . . . They said they don't have any indications of that. They believe, which would be true, there will be screaming radiation levels if those walls were knocked out. They said the walls . . . could have been knocked out, but the spent-fuel pool wall may not be the outermost wall."

The Japanese had shown Monninger a "bird's eye schematic" of the Unit 4 spent fuel floor. "They had a decent level of confidence that . . . the spent-fuel pool walls hadn't been blown out and the success path was to flood that up and not to go with sand, dirt, or whatever."

Borchardt called again from Capitol Hill looking for any new information. Monninger's report from the Japanese was repeated: "They don't think the spent fuel pool for Unit 4 is as degraded as we thought." Borchardt asked for more details. Cooling in the pool still may be possible, he was told. Monninger added that he believed reactor fuel had been ejected in the explosion at Unit 4 and was now lying around the plant site, which would explain high radiation readings on the ground. The Japanese were bulldozing dirt over the rubble, and the radiation levels dropped by as much as 70 percent.

"So the potential is solely to refill the pool?" Borchardt asked.

"That's correct," came the reply. Even so, Borchardt was told, "there's no change to the recommendations that we've made." The fifty-mile evacuation stood. As best as the White Flint crew could tell, Units 1 to 3 had damaged cores and the spent fuel pool at Unit 4 was almost dry.

The NRC had sound reasons for its suspicions. There was no clear proof, but there was plenty of circumstantial evidence. The Unit 4 pool contained the most, and the hottest, spent fuel among the reactors at Fukushima. It would therefore be the first to boil after cooling was lost. The Unit 4 reactor building had blown apart, as TV replays nearly constantly reminded them; the most likely reason would be the detonation of hydrogen gas. Since the Unit 4 reactor core held no fuel, the most likely source of hydrogen gas was exposed fuel in the pool. If the fuel were uncovered, its metal cladding could ignite—explaining the fire reported in Unit 4. And the high radiation levels reported between the Units 3 and 4 buildings could be caused by debris, perhaps even portions of fuel rods, ejected from a dry Unit 4 pool during the explosion.

Finally, observers reported white vapor emanating from the Unit 3 reactor building while nothing visible wafted from Unit 4. Based on this latest circumstantial evidence, it was reasonable to conclude that the Unit 4 spent fuel pool had boiled or drained itself dry while the Unit 3 pool was boiling toward that outcome.

But still, there was no definitive proof. The Japanese could turn out to be right after all, but it would be many days before anyone could get close enough to Unit 4 to find out.

• • •

Jaczko had been scheduled to discuss the NRC's 2012 budget with two House
Energy and Commerce subcommittees, but Fukushima was on everyone's
mind. Before Jaczko could testify, he was summoned to a meeting at the White
House. Jaczko told President Obama and his national security advisors that
the NRC recommended the evacuation of Americans living within fifty miles
of Fukushima Daiichi. It was, the chairman said, the advice the NRC would
give if this incident were taking place in the United States.

NRC chairman Gregory Jaczko tells a House subcommittee hearing on March 16 that the
NRC believed the spent fuel pool at Unit 4 was dry, raising the possibility that the spent
fuel could catch fire and release additional radioactive material. The announcement came
shortly after the United States advised its citizens living within fifty miles of Fukushima
Daiichi to evacuate or remain indoors if evacuation were not possible. *U.S. Nuclear Regulatory Commission*

A short time later, White House Press Secretary Jay Carney alerted reporters to the expanded evacuation recommendation during a briefing. The
State Department had issued a travel warning half an hour before, citing the
"deteriorating situation" at Fukushima Daiichi. U.S. citizens living within fifty
miles (eighty kilometers) of Fukushima were advised to leave or take shelter
indoors if evacuation was not possible. In addition, the warning said, "the

State Department strongly urges U.S. citizens to defer travel to Japan at this time and those in Japan should consider departing."

Washington's fifty-mile evacuation recommendation was about to get some competition in the news. Before he began his testimony on Capitol Hill that afternoon, Jaczko called in to the White Flint operations center. "[J]ust to repeat, we believe pool No. 4 is dry, and we believe one of the other pools is potentially structurally damaged?" he asked. "That's correct," Chuck Casto said from Tokyo. "That's the best we know."

"I mean the relevant factor is it's dry," said Jaczko. "Yes," Casto confirmed, "and they can't maintain [water] inventory at all."

Jaczko hung up, but the conversation continued between the NRC experts in Tokyo and White Flint. They realized they had neglected to tell the chairman that the Japanese didn't agree with the NRC's assessment about the Unit 4 pool. They didn't call him back, though, because they thought he didn't want that level of detail. And there were plenty of other problems to worry about. The Unit 2 spent fuel pool now appeared to be in trouble; steam was rising through the hole created in that reactor building's wall by the Unit 1 explosion, so that pool might be boiling dry. Steam or smoke was coming from the Unit 3 pool, meaning the water level there could be dropping. Unit 1 appeared to have the only pool of the four not in immediate trouble.

When Jaczko took his seat in the large House hearing room, he launched into the latest news from Fukushima. At three reactors there is "some degree of core damage from insufficient cooling," he said. At Unit 2, cooling was not stable but the primary containment was functioning. Water levels in the Unit 2 spent fuel pool were decreasing. The integrity of the Unit 3 pool had been compromised.

Then he moved on to Unit 4. "We believe that secondary containment has been destroyed and there is no water in the spent fuel pool," he said, "and we believe radiation levels are extremely high, which could possibly impact the ability to take corrective measures." Jaczko said the highest priority now was to maintain cooling of the reactors at Units 1, 2, and 3 and to keep water levels up in the spent fuel pools. If cooling functions should fail, he warned, "it would be very difficult for emergency workers to get into the site and perform emergency actions."

What poses the highest risk at the moment, asked Rep. Henry Waxman, a Californian with a long involvement in nuclear safety issues.

"All of the factors together, really, the combination," Jaczko said. "There's the possibility of this progressing further." If the cooling systems failed, he explained, emergency workers "could experience potentially lethal doses in a very short period of time."

Jaczko hustled out of the House hearing and across Capitol Hill to the Senate, where Bill Borchardt had been pinch-hitting for his boss before the Senate Environment and Public Works Committee, which had convened specifically for a briefing on Fukushima. The chairman was grilled about the safety of U.S. reactors, especially those in seismically active areas, such as the two in the home state of the committee's chair, California senator Barbara Boxer.

After Jaczko's testimony, the committee heard from representatives of the Union of Concerned Scientists and the Nuclear Energy Institute (NEI), the primary nuclear industry lobbying group in Washington. They were asked to address whether a Fukushima-type accident could happen in the United States and if so, what needed to be done to preclude such a disaster.

Tony Pietrangelo, chief nuclear officer of the institute, chose his words carefully: "I think I understand your concern, because I share it, that people are seeing what's happening in Japan and they're scared. We can never say that that could never happen here. There's no such thing as a probability of zero. . . . But what I would tell you is it doesn't matter how you get there, whether it's a hurricane, whether it's a tsunami, whether it's a seismic event, whether it's a terrorist attack, whether it's a cyberattack, whether it's operator error or some other failure in the plant, it doesn't matter. We have to be prepared to deal with those events."

Edwin Lyman of the Union of Concerned Scientists agreed that a Fukushima-type event could not be ruled out in the United States: "We have plants that are just as old. We have had a station blackout. We have a regulatory system that is not clearly superior to that of the Japanese. We have had extreme weather events that exceeded our expectations and defeated our emergency planning measure[s], [such as] Hurricane Katrina."

While Pietrangelo said that the industry *had* to be prepared to deal with severe events like natural disasters, he did not say that the industry actually *was* prepared to deal with them. In fact, he told the committee that Fukushima had prompted an industry-wide effort to "verify each company's capability to mitigate conditions that result from severe adverse events," including total loss of electric power, earthquakes, and floods. At the time, he did not anticipate that those reviews would turn up numerous problems, calling into question the level of preparedness of U.S. reactors.

Pietrangelo informed Boxer that "we already have mobile—diesel-driven mobile pumps on every site in the country that can be moved around the site to provide another contingency measure should we lose a cooling source. And there's countless other measures like that."

Those measures are known as "B.5.b," named after the section of the regulatory orders where they appear.[4] They were among the steps that the NRC required utilities to take after the 9/11 attacks to prepare for "loss of large areas of a plant due to explosions or fire"—the kind of damage that might be expected from a terrorist aircraft attack. For example, utilities must store B.5.b equipment far enough away that it might be able to survive a plane attack on a reactor. But there is no requirement that the equipment survive other types of disasters, such as an earthquake. When Lyman asked about the ability of the emergency pumps to survive an earthquake, Pietrangelo said that the pumps are not certified to withstand any earthquake at all, much less a severe one.

In other words, in the face of a Fukushima-scale event, the B.5.b measures could well be worthless. But in the days after Fukushima, both the industry and the NRC cited the B.5.b measures in response to the question they had both feared: "Can it happen here?"

Jaczko's assertion that the Unit 4 spent fuel pool was dry apparently didn't sit well with the Japanese. Casto and Tony Ulses were invited to the Kantei, the prime minister's official residence and office. The Japanese wanted to show the Americans a video of the Unit 4 spent fuel pool—indicating it contained water. Monninger told the White Flint crew: "We think the reason they're doing that is because, I guess, of maybe statements the NRC has made or maybe the chairman's hearing testimony . . . saying that Unit 4 the spent fuel pool was dry."

A short time later, Casto reported back after watching the video, which the Japanese declined to provide to the Americans. "You know, it's not very clear. You're talking about a helicopter that's trying to do a lot of things at once in a field. And they tried to scan all four [reactor] units. You have to look through a window. And they claim there's a reflection of water on the Unit 4 spent-fuel pool. . . . There's something there. You don't know if it's steel or water. They claim it's a reflection."

Steam was visible coming from the side of the building nearest the spent fuel pool, said Casto, obviously frustrated. "You've got a building that's had an explosion and has debris everywhere, and you're trying to look at it with a helicopter that's flying by in split seconds. You can't tell anything in there. You know, they claim there's a glimmer of a reflection but, you know, it's steaming. Unit 3 is steaming even harder."

Jaczko, who had joined the call, interrupted Casto. "So, at this point, you no longer believe that the pool is dry?" Casto replied: "I would say, as of five o'clock yesterday, the pool had some water in it."

"Okay. Now I've said publicly the pool is dry," said Jaczko. "Do you think that's inaccurate?" Casto replied: "I would say it's probably inaccurate to say it's dry."

"Do you think I need to roll back any of the statements that I made?" Jaczko asked.

"I don't think so," Casto said. "It may not have been dry, but it certainly wasn't full." There was no need to change the fifty-mile evacuation advisory, the team assured Jaczko.

As debate swirled at White Flint, President Obama and Prime Minister Kan were on the phone. Obama promised continued technical assistance to Japan and the two leaders discussed the steps the United States was taking to protect its citizens. After the conversation, Japanese cabinet secretary Yukio Edano told reporters in Tokyo that the two countries would do a better job of sharing information.

As for the Americans, everyone understood that the United States needed to tread lightly. As Chuck Casto later observed of the situation: "The best science person can't bring science to it if they don't have the right diplomacy skills."

Convinced that the Unit 4 spent fuel pool held water, the Japanese focused their attention on Unit 3. The white smoke seen billowing from that unit the day before—as well as a spike in radiation—made getting water to the fuel there a priority. A three-pronged attack began shortly before 10:00 a.m. on Thursday, March 17.

It consisted of dropping water from Self-Defense Forces helicopters and spraying water from the ground using SDF fire trucks and high-pressure water cannon trucks that the Tokyo Metropolitan Police Department used for riot control.[5]

An initial attempt to dump water from an SDF helicopter on March 16 had been abandoned because of high radiation levels. Lead plates lined the bellies of the helicopters and the crews wore protective clothing when they took off the next day.

Over the course of ten minutes, two CH-47 Chinook helicopters dropped thirty tons of water on the upper parts of the Unit 3 reactor building during four flights. NHK television cameras, positioned about twenty miles from the plant, captured the drops in footage viewed around the world. Although a small amount of white steam rose from the building, very little water apparently reached the pool. Most was carried off by strong winds, and some was deflected by debris atop the pool. The water cannons were useless and the fire trucks proved inadequate; neither could deliver the water where it was needed. Radiation levels prohibited moving the equipment closer.

TEPCO also focused on restoring power to the plant via a new transmission line, although utility officials were unable to say when that might be completed. The plan called for repowering Unit 2 to restore cooling to its spent fuel pool. Because Unit 2's reactor building was intact aside from a small hole, it would be difficult to spray water inside (in contrast to Units 1, 3, and 4, which no longer had roofs). But given the extensive damage to the plant's wiring systems, there was no assurance that off-site power would be of much use.

The first radiation readings gathered by the U.S. National Nuclear Security Administration's Aerial Measuring System, part of the DOE's Radiological Assistance Program, were providing independent data. Until now, most of the measurements had come from TEPCO via NISA, and Washington wanted its own. The U.S. sensors not only measured how much radiation was present, but also identified the isotopes emitting it—the nuclear fingerprints of dangerous materials including cesium-137. In addition, a U.S. Air Force Global Hawk drone flew over the plant site, measuring temperatures and gathering high-resolution imagery.

While the Japanese struggled to get water into the pools, the Americans were offering five heavy-duty pumps from the U.S. Defense Threat Reduction Agency that were available for pickup three hours away from the plant. (Earlier, the Japanese had rejected two fire trucks offered by the U.S. Air Force because the vehicles were not registered in Japan and thus could not be legally driven on the roadways, an act of bureaucratic nitpicking that amazed the Americans.) The pumps could deliver seawater to the fuel pools—assuming the pools were still intact. That worried Casto. "The pumping strategy may not be useful at all if there is no spent fuel pool, and there is just a rubble bed in there somewhere," he said.

Even so, it seemed a logical plan of attack. At 6:00 a.m. on March 17, the Japanese SDF picked up the pumps at Yokota Air Base and set out for Fukushima Daiichi, 196 miles away.

• • •

Delivering the water would require more than pumps, however, and U.S. experts had spent hours conferring on just what was needed. Pipes and strainers (to keep debris in the ocean water from clogging the pumps or the spray nozzles) had to be acquired. Because of the extremely high radiation, installation of the system had to be completed as rapidly as possible. "[I]t's got to be on tractor-trailers, minimum disconnect," one member of the U.S. team said. "They would drive in, hook this thing up, throw the portable submersible pump out in the ocean, and light this system off and leave." The NRC asked the giant engineering firm Bechtel Corporation to design the complete cooling systems that would be needed and to arrange for procurement and delivery of all the additional equipment to the site, but not to perform the dangerous jobs of installing and operating the systems, which would be left to the Japanese.

The NRC was developing a radiation dose map of the plant site to help installers minimize their exposure. "[I]t's going to be heroic efforts, because you're going to have somebody running toward the pool," said one of the White Flint staffers. Boron would be added to the water to reduce the likelihood of the spent fuel going critical. Twenty thousand pounds of boron had been located at California's Diablo Canyon nuclear plant and was being flown from Vandenberg Air Force Base to Yokota.

An exhausted Chuck Casto called in to White Flint to say the pumps had arrived at the plant and efforts were now under way to reduce the radiation levels with sand or lead shielding so the system could be installed. "[Y]ou know there are lethal dose rates they're getting outside that building," said Casto.

The discussion turned to Unit 2, and Casto was asked about its status: had the primary containment been breached? He did not know. A NISA summary was vague, Casto said, referring only to the possibility of an "incident" in the suppression chamber, based on the sound of an explosion and a subsequent reduction in the measured pressure in the torus.

For a moment, the magnitude of the crisis for the Japanese overwhelmed Casto, who noted that a thousand bodies had washed ashore from the tsunami. "[I]t's hell over here for that government. I mean, it's just absolutely hell. And I know we get frustrated with them, but, man, when you think about what they're faced with, it's absolutely unfathomable."

Although the United States was working out details on the pump system, it was a one-sided effort. "We are not getting any takers from the Japanese side of this equation," Virgilio told his colleagues. Although the SDF had collected the pumps and delivered them to the plant, the Japanese apparently were afraid that trying to hook up the system would be too dangerous because of the high radiation levels. So they were intent on continuing their spraying operations. "They are using helicopters the same way you fight forest fires," said a member of the American team.

At about 8:00 a.m. Thursday, Dan Dorman, on duty at White Flint, briefed the NRC team via a conference call. Events had moved both Unit 4 and Unit 2 down the list of concerns. The highest priority now was the Unit 3 spent fuel pool, which appeared to have experienced a zirconium-water reaction and a resulting fire, he said. Over the previous twelve hours, the Japanese had tried to drop and spray water into the Unit 3 pool, but without much effect. Seawater injection to the reactor vessels continued in Units 1, 2, and 3, with water levels in the cores at about half the height of the fuel. There were mixed reports on the status of the Units 2 and 3 containments. TEPCO was still trying to restore power to the site. The pumps were in place but it was not certain that they could develop enough pressure to spray water high enough to reach the pools. New radiation measurements had been collected by aircraft overnight and were being analyzed. Radiation levels of 375 rem per hour were measured three hundred feet above the Unit 3 reactor.

To everyone's surprise, the Japanese asked the NRC to list its priorities on controlling the situation at Fukushima Daiichi. Although the turnabout was welcome, fulfilling the request would put the NRC in the position of directly offering advice, something it had thus far avoided. Making a mistake weighed on Casto. "I don't want to try any strategy that doesn't have sound measures, or reasonably sound measures, of success," he told Virgilio. "Because the last thing we want to do is make the situation worse for the Japanese. We're going to own this thing if we do."

Radiation exposure fears were growing beyond the vicinity of Fukushima Daiichi. The NRC was aware that significant levels of fallout could occur outside of the fifty-mile evacuation zone. For data on that, it had to depend on the DOE, which was using its more sophisticated atmospheric modeling tools to develop dose projections extending one hundred and fifty to two hundred miles from the plant, an area that included Tokyo. The DOE was also running simulations to estimate possible doses to the public in Alaska, Hawaii, and the West Coast, and it was getting some alarming results: for instance, doses as high as thirty-five rem to the thyroids of people in Alaska. The NRC didn't trust the numbers the DOE was coming up with, but it didn't have the means to disprove them.

In the United States, the Department of Homeland Security was screening passengers returning from Japan. In China, grocery stores in several cities were stripped of iodized salt after rumors spread that a radioactive cloud was moving in. Some buyers mistakenly believed that iodized salt would protect against radiation; others feared that sea salt supplies from the coast would be contaminated. The city of Florence, Italy, sent a plane to Japan to pick up its stranded orchestra and the conductor, Zubin Mehta. (Maestro Mehta would return a month later for a performance to benefit disaster survivors.)

Casto and the NRC squad in Tokyo were swamped with meetings—with the ambassador and embassy staff, with other U.S. officials aiding in the American response, with Japanese government officials. The crew was running on six hours of sleep or less, with sessions stretching well into the early morning hours. Noticeably absent, however, were representatives of TEPCO. The utility seemed intent on going it alone.

Finally, TEPCO officials did agree to sit down with the Americans to discuss a pumping system the United States had proposed for cooling the spent fuel pools. It was the first time the utility had heard of the U.S. plan, Monninger told his colleagues at White Flint. Apparently NISA and METI, which had been involved in the discussions of the U.S. plan, had not bothered to apprise TEPCO.

Yet the United States had already given Bechtel the go-ahead to assemble four separate pumping systems, or "trains," one for each of the fuel pools. By March 18, the components had been gathered in Australia, awaiting an airlift to Japan. Only then did the U.S. team discover that the Tokyo Fire Department had assembled a similar pumping system with local parts for use at Fukushima Daiichi.

No matter whose system would be used, the problem, of course, was the final leg: getting close enough to lay water lines near the reactors and pools. Radiation levels between Units 2 and 3 now ranged between 450 and 600 rem per hour.

TEPCO's preferred and potentially less risky approach was to use two concrete pumping trucks, similar to those used to encase the Chernobyl reactor after that accident. The trucks, equipped with booms capable of extending about two hundred feet, could remotely spray seawater into the fuel pools with precision. Two such trucks were in Japan, and additional ones could be obtained from abroad.

Casto was still worrying about pumping water onto the hot fuel. The risk was a massive radioactive steam cloud "that's going to . . . panic everybody."

"We're in such never-never land, and I don't really know how to make decisions when you have very little information," Casto said, and urged his colleagues to seek additional guidance from experts inside and outside the government. "We're going to need all the great minds together," he added. "Let's solve this as an American nuclear industry helping the Japanese nuclear industry, not just the NRC."

HELPFUL HINTS

As the experts in Japan and the United States struggled to find ways to cool the cores and stem the radiation releases spilling from Fukushima Daiichi, average Americans had no shortage of ideas—nor any reluctance to share them. They flooded the NRC with suggestions via e-mails and phone calls.

"I'm sure y'all have many experts trying to assist in the emergency in Japan," wrote Sharon H., "but I thought I would write just in case this idea had not been considered": cornstarch.

"Have you considered . . . fine ground black pepper to seal the crack at Fukushima Daiichi?" wrote Shawn M. "Yes, common household pepper is very effective as a crack sealant. I once sealed a long crack (8") in an engine block. It held for years and I did not need to replace the engine."

Among other suggestions: Ping-Pong balls, antifreeze, a ski resort snowmaker to blow snow into the core. Coal ash. An asbestos blanket. Liquid nitrogen. Putty.

Vinny S. proposed using tiles from the space shuttle. "The shuttle program is ending and we could recycle materials and be eco-friendly while ensuring safety at the reactors."

Others were more insistent: "Get this information to Japan or tell them when they call you," directed one e-mail. "I was divinely led. God is telling me it will work." What followed was a detailed list of instructions entailing "vector forces," a "metrogravitron particulator," and a device known as the "Deometrian."

Most suggestions received a standard response from the NRC's Office of Public Affairs: "We appreciate suggestions that work toward resolving the situation in Japan; it's reassuring to see how helpful and dedicated private citizens have been in light of this disaster. . . . Please understand that the NRC has some of the most expert people in the world available to assist the Japanese authorities in whatever way they request." The NRC suggested the Americans also contact the Japan Atomic Energy Commission.

One public-spirited citizen apparently got a runaround before finally landing at the Office of Public Affairs. His name was Harold Denton. He's "been calling all over NRC," wrote staffer Amy Bonaccorso. In case the name didn't ring a bell, she added that Denton was the "spokesperson for the NRC during Three Mile Island—he also was the person

[who] provided daily reports to Pres. Jimmy Carter." Denton, who also delivered daily briefings to hundreds of reporters during the Three Mile Island accident, wanted to urge the commission to make its experts available to the media, she wrote, because reporters "are reaching out everywhere for info . . . including retired employees (with old information)." Bonaccorso's boss promised to call Denton.

Stepping into the information void on March 15 was TV personality Glenn Beck. "Glenn Beck is now explaining the China Syndrome, using a wok as a containment vessel," an Office of Public Affairs staffer e-mailed his colleagues, to which one responded: "ohmigod." "Maybe it's a big mixing bowl, I can't tell," said a follow-up e-mail. That wasn't Beck's only prop: M&Ms made handy fuel pellets.

That collaboration was already in the works. The next day, Saturday, March 19, a group of experts from industry met with their government counterparts at White Flint. The model, Virgilio explained, was the public-private response to the Deepwater Horizon oil spill in the Gulf of Mexico in 2010. Attending were representatives from the Institute for Nuclear Power Operations, which was organizing the meeting, GE, Exelon, the French nuclear conglomerate Areva, the DOE, and the Electric Power Research Institute. Experts within the nuclear industry had regularly been conferring with the NRC and Japanese authorities as well as conducting their own research. The Americans hoped that by merging the private and public response, TEPCO might become more receptive to outside help.

The Japanese government might also need some convincing. "We're going to start working through diplomatic channels in-country to try to make sure that what we develop is implementable from a political standpoint," said Virgilio. Key to any public or private response was reliable data. And that was still in short supply.

5

INTERLUDE—SEARCHING FOR ANSWERS: "PEOPLE . . . ARE REACHING THE LIMIT OF ANXIETY AND ANGER"

From the first days of the accident, Yukio Edano, a forty-six-year-old lawyer, became the face of the government. As chief cabinet secretary, responsible for coordinating the executive ministries and agencies, he was omnipresent on live TV.[1]

Although Edano gained certain celebrity status, his carefully worded statements later came in for criticism. Even as he announced evacuations, he failed to adequately explain to the public the reason for the orders; other important details often were missing from his summaries.

As the accident unfolded, the briefings of various government agencies, as well as the updates provided by TEPCO and NISA, other government ministries, and local governments, were a disjointed mix of reassurances and frustratingly vague descriptions of what was taking place at the plant. At times, the authorities provided conflicting information.

By March 18, official responses to the accident seemed to become a numbers game, with statistics bandied about in a confusing blur. That day, NISA elevated the ranking of the Fukushima Daiichi accident from a level 4 to a level 5 on the seven-level International Nuclear and Radiological Event Scale (INES) because of fuel damage to the cores at Units 1, 2, and 3. In a press release announcing the change, TEPCO blamed the reactor crisis on "the marvels of nature," a reference to the tsunami and earthquake.

While the significance of that increase from level 4 to level 5 undoubtedly escaped many in the public, for regulators and the Japanese nuclear industry it was as much a symbolic as a substantive change. Japan's worst nuclear accident to date, the 1999 criticality incident at the Tokai fuel fabrication facility that killed two workers, had rated a level 4 on INES. Three Mile Island was a level 5, and now Fukushima Daiichi was joining Three Mile Island on that dubious global list.

What Tokyo and the Japanese nuclear industry dreaded even more was having to acknowledge that the accident warranted a level 7 rating, defined as a major accident with widespread health and environmental effects and the external release of a significant fraction of reactor core inventory. Only one accident in history had been deemed that serious: Chernobyl.

It would be almost four more weeks before Japanese authorities announced—belatedly, according to many—that the events in Fukushima indeed rated a level 7 ranking. In the annals of nuclear history, Fukushima Daiichi and Chernobyl would forever be linked.

Edano's news briefings routinely contained two messages: the latest radiation readings posed no "immediate risk to health," followed by these words: "Remain calm." The Japanese heard that advice from their government frequently during March as radioactive plumes spilled from the plant, as the expanding danger zone forced evacuees to move repeatedly, and as authorities continually failed to contain—or even explain—the accident. If more evidence was needed that things were out of control, videos of the explosions and damage at Fukushima Daiichi played repeatedly on TV and websites, occasionally serving as background illustrations for newsreaders repeating the government's reassuring statements.

Japan's chief cabinet secretary Yukio Edano was the government's primary spokesperson during the accident. Frequently advising the public to remain calm, he later came under criticism by investigators for downplaying the seriousness of threats posed by events at Fukushima Daiichi. *AP*

Such assurances had begun to ring hollow. "People in Fukushima are reaching the limit of anxiety and anger," the governor of Fukushima Prefecture complained a week into the accident. They weren't the only ones. Across Japan, millions of people struggled to make sense of what was happening, and many began looking for alternative sources of information.

Criticism was also mounting internationally about Japan's candor regarding conditions at the plant and the degree of risk posted by radiation emissions. This clearly had grown beyond a domestic event. Early on, fears that dangerous levels of radiation could contaminate areas far beyond Japan gained legitimacy in part because of the dearth of information. The fragments of data experts could glean from Fukushima Daiichi painted a picture often at odds with the assessments coming out of Tokyo. The NRC's experts undoubtedly were not the only ones abroad engaged in debates about the potential radiation hazard.

The International Atomic Energy Agency was faulted for its sluggish and confusing response. The agency's head, Yukiya Amano, a career Japanese diplomat, paid a one-day visit to Tokyo on March 18 to announce that the IAEA and the international community were "standing by Japan." Critics argued that instead of standing by, the IAEA should be wading in and participating much more directly in efforts to assess the threats posed by the reactors. Some blamed the organization's backseat role in part on Amano's ties to Tokyo's political and nuclear establishment. (That criticism increased when the IAEA declared in late May that Japan's response to the accident had been "exemplary.")

However, even if the IAEA could have done more, it was hamstrung by its lack of authority to intervene in the internal nuclear safety affairs of sovereign states. With regard to civilian nuclear power safety, the organization functioned more to set standards and practices than to be a nuclear cop on the beat. And at the beginning, the IAEA scientists were as starved for information as those everywhere else outside the Japanese government and TEPCO.

Logically, many Japanese turned to the news media for help. There, too, they were often ill served. For journalists, the events at Fukushima Daiichi posed unique reporting challenges. The accident superimposed a complex technological failure involving multiple reactors on a catastrophic natural disaster that itself was a major story. Nuclear jargon was confusing and unfamiliar; radiation measurements were baffling. Information came secondhand; access to the plant and those working inside was impossible.

Reporters from Japan's major media outlets stayed well outside the twelve-mile (twenty-kilometer) evacuation zone, not traveling closer for more than a month after the accident. Rather than do on-scene reporting, journalists often just repeated government and utility statements, supplementing them with interviews of academics or industry spokesmen, whose objectivity soon came into question.

Some foreign journalists and Japanese freelance reporters ventured as close to the disaster scene as they could go, talking with evacuees and providing vivid descriptions of the confusion and miscues in the accident response. At times their accounts were so at odds with what major Japanese media were reporting that these journalists were accused of sensationalizing.

The international and independent reports were often posted or streamed on websites. For those with Internet access who saw the accounts, the events at Fukushima Daiichi seemed far more alarming than how they were portrayed in regular updates from official sources.

Advocacy groups, including the Citizens' Nuclear Information Center, quickly became sought-after providers of trusted information. On March 12, the center began conducting extensive daily briefings in Japanese and in English, delivered by independent nuclear experts. Streamed live on the Internet, the sessions provided details and analysis often unavailable elsewhere. They quickly developed a large following, with viewers e-mailing questions for the experts to answer. International organizations, including the Union of Concerned Scientists, also conducted detailed briefings, posting transcripts and relevant documents online and responding to hundreds of media inquiries from the United States, Japan, and elsewhere.

As the days progressed, and the hunger for information grew, new media outlets—the Internet, Twitter, blogs, Web-based broadcast media—played an increasingly vital role. Some of the content for these outlets came from those most immediately affected: disaster victims, evacuees, and even emergency workers who captured events on cell phones or reported them via text messages, blogs, and Twitter feeds. Again, the stories they told were often at odds with the official assessments.

One of the most compelling accounts came from Katsunobu Sakurai, the mayor of Minamisoma, a devastated coastal community about fifteen miles from Fukushima Daiichi. Sakurai sat down in front of a camcorder in his office and pleaded for assistance from anyone. The natural disaster and the nuclear accident had made conditions in his city untenable, he said. Food and fuel deliveries had stopped; residents who had not already fled had been ordered to remain indoors. "With the scarce information we can gather from the government or TEPCO, we are left isolated," Sakurai said, looking into the camera. "I beg you to help us. . . . Helping each other is what makes us human being[s]." The eleven-minute recording, posted on YouTube, was viewed by more than two hundred thousand people in the course of three weeks, and relief poured into the beleaguered city.

Even as up-to-date technology made disseminating the news faster and simpler, many Japanese journalists labored under the influence of traditional politics, economics, and culture, which did not reward confrontation.

As a result, TEPCO often got a free ride from potential critics in and out of government, and it grew to symbolize much of what was wrong in corporate management and regulatory oversight in Japan. Despite its well-known history of covering up safety problems, the utility was regarded by many as a "cornerstone of corporate Japan," relying on political muscle, public goodwill, and deferential watchdogs to keep its nuclear facilities and the billions invested in them operating without serious challenge. TEPCO had committed to building more reactors at home and overseas, and was even planning to help fund a two-reactor project in Texas. What confidence could reporters have that the company was now being forthright?

In the absence of company president Masataka Shimizu, who had vanished from public view, the utility put junior executives before the cameras to apologize for "causing inconvenience." But they offered little in the way of new details. TEPCO's press releases occasionally omitted crucial information. The utility announced that it had successfully begun injecting seawater into the Unit 2 reactor at 11:00 p.m. on March 14, for example, but failed to mention that radiation levels had jumped at the plant entrance about four hours earlier.

Members of the media camped out for days at TEPCO headquarters, squeezed into a cluttered room off the first floor lobby. Utility officials would appear to make brief announcements at all hours. When pressed by reporters for details and explanations, they often were unable or unwilling to answer.

In better times, TEPCO had been a savvy corporate communicator with deep pockets and enviable national clout. For years the company had financed a sophisticated and expensive public relations program to promote the benefits—and safety—of nuclear power. TEPCO, like Japan's other nuclear utilities, erected elaborate visitor centers that resembled theme parks, filled with animated characters extolling the wonders of nuclear power. TEPCO's mascot, Denko-chan, promoted the company to the younger set and their families.

TEPCO ranked alongside Japan's internationally known corporations—Panasonic, Toyota, Sharp—in dollars invested for advertising. But when things went awry, TEPCO seemed to have no crisis communications plan in place, just as it lacked a workable accident plan.

As for government officials, they also had incentives to downplay the accident. Japan had pinned its energy future on nuclear power. Every journalist was aware of the historically close ties between industry and government. What confidence could reporters have that public officials would now provide crucial information that reflected poorly on such an influential industry—or on the policy decisions that had promoted its expansion?

Public trust was an early casualty. Yet trust is critical, noted a Diet-appointed committee, led by Yotaro Hatamura, that investigated the crisis and detailed its findings in a lengthy report. The Investigation Committee on the Accident at the Fukushima Nuclear Power Stations of Tokyo Electric Power Company wrote, "Inappropriate provision of information can lead to unnecessary fear among the nation."

The report echoed the findings of another independent investigative commission, which criticized the government for withholding information on the basis that it had yet to be completely verified. By "sacrificing 'promptness' in order to ensure 'accuracy,' there is conversely a danger of inviting citizen's [sic] mistrust and concern," the committee noted, acknowledging the difficult balancing act that often faces officials in times of emergency. That was particularly true when news was flooding in from multiple sources other than the government.

Perhaps the greatest chasm between what was being said and what the public needed to hear concerned the complex issue of radiation exposure and health risks. The looming question on everyone's mind was the obvious one: are we safe? There could be no unqualified answer; a "safe" level of radiation is an issue on which scientists and health experts often disagree. Even so, the government failed to provide basic guidance, a criticism leveled by the Hatamura committee.

Chief Cabinet Secretary Edano and others repeatedly used the ambiguous phrase "it does not have immediate effects on health." As the Hatamura investigation noted, that expression could be interpreted in two conflicting ways: that the radiation was harmless, or that it might produce cancer, just not right away. Citizens were left to ferret out the truth on their own.

Time and again, the government bungled its handling of the radiation and health issue. On March 16, for example, as the United States was ordering a fifty-mile evacuation for its own citizens in view of rising radiation levels, Edano declared that only the government's Nuclear Safety Commission could provide an accurate analysis of radiation data. But the chairman of the NSC, Haruki Madarame, was occupied advising Prime Minister Kan and apparently unavailable. On March 23, nearly two weeks into the accident, Madarame held his first news briefing and informed reporters that no analyses were available "because we are very understaffed."

By then, citizen activists, armed with their own radiation monitoring equipment, had begun to gather readings around the country and post them online. The flurry of numbers—some reliable, some unverifiable—only compounded the confusion. For parents desperate to protect their children, trustworthy answers were hard to find.

A child receives radiation screening at a gymnasium in Fukushima on March 24, 2011. *AP*

In an effort to avoid arousing fears, the government also deliberately withheld crucial information—a fact that confirmed the suspicions and inflamed the distrust of many when the omissions came to light.

It wasn't until June, for example, that officials publicly acknowledged meltdowns at the three reactors—information the government had possessed since March 12, based on the off-site detection of tellurium-132, a fission product that could only have come from a melted core. By June it was old news to anyone in Japan who had Internet access and cared to search for Fukushima analysis from other sources. Moreover, the Japanese had acknowledged on April 12 that Fukushima Daiichi was a level 7 accident on INES, a ranking that certainly implied a meltdown. The official disclosure of the meltdowns finally arrived shortly before details about them were to be released at an international conference.

And there was the Japanese media itself. It, too, operated under a cloud. Reports about Fukushima Daiichi frequently came colored with the built-in biases of the publication or broadcast outlet: for or against nuclear power, for or against the political party in office, for or against Japan's powerful but faceless elite who influenced policy but seemed unaccountable for their actions.

Japan's news media for decades have functioned through a system of press clubs whose "members"—the journalists from mainstream media—work from newsrooms inside government offices and have ready access to the public officials and agencies they are assigned to cover. Nonmembers, who include reporters working for the growing number of Web-based or independent news organizations as well as freelance journalists, are left to their own devices, briefed less frequently and otherwise isolated from the daily official media loop. (Foreign journalists have their own press club, which accords them certain privileges, but they also are generally less constrained in their relationships with public and private newsmakers.)

Critics of the press club system have long argued that it fosters a too-cozy relationship between reporters and those they cover. Journalists can be reluctant to challenge the official line for fear of alienating sources and losing access. Politicians and bureaucrats, for their part, know their pronouncements will be duly reported without countervailing views because that is what's expected of Japanese press club members. Aggressive investigative reporting, common in many other countries, has been the exception rather than the rule for many years in Japan.

The press clubs also can limit coverage of those who take issue with government policies. Japanese antinuclear activist Aileen Mioko Smith, who heads Kyoto-based Green Action, described the difficulty of attracting media attention to nuclear safety issues. "The Japanese press club system has proven very effective at keeping nuclear news *out* of mainstream media," she wrote shortly after the accident. "I often say, 'If you want to make sure that you don't get any media coverage, go to the METI press club.' " (METI, the Ministry of Economy, Trade and Industry, oversaw nuclear safety until a government reorganization after the Fukushima Daiichi accident.)

And thus, in the frantic days following March 11, when information became the coin of the realm, millions of Japanese felt themselves abandoned. The institutions they had long depended on and trusted—government, corporations, the media—now seemed ineffectual, even suspect, failing to deliver as a result of intent, ineptitude, or the sheer immensity of the unfolding crisis.

The experience left its mark on many Japanese. "There is a sense of betrayal," says Dr. Evelyn Bromet, a psychiatric epidemiologist at the University of New York at Stony Brook whose research includes the mental health impacts of the Three Mile Island, Chernobyl, and Fukushima accidents. She visited Fukushima and examined detailed survey responses collected from evacuees. The responses reveal a high level of fear, anger, and emotional stress among many Japanese, she says. "There's nobody that they trust any more for information."

"You shouldn't assume that people can't handle the truth," says Bromet. "It may be difficult to swallow, but it's better to be open and straight with them."

During the Fukushima Daiichi accident, Japanese authorities ignored that basic tenet of crisis management, concluded Japan's Nuclear Accident Independent Investigation Commission in its July 2012 report. "[T]he government chose to release information purely from a subjective perspective, rather than reacting to the needs of the public."

Just as the NRC's technical staff rapidly geared up in response to the events at Fukushima Daiichi on March 11, so did the commission's public information arm. Although the accident was taking place halfway around the world, U.S. media would have questions.

As the hours passed and details trickled out of Japan, it became increasingly clear that these distant events could have major implications for American citizens abroad, and even at home, if radiation releases became severe enough. In addition to those issues, the public affairs staff also was sensitive to possible repercussions for the U.S. nuclear establishment, only now getting back on its feet after a lengthy, involuntary hibernation.

Construction had resumed on the first new American reactor in decades, and the NRC was well on its way toward approving combined construction and operating licenses for four more reactors. The White House supported the expansion of nuclear power in the U.S. energy mix, and the NRC was systematically extending the licenses for aging plants, giving them another twenty years of operation. Some were even calling this a nuclear "renaissance."

Now, however, opponents of the growing footprint of nuclear power in the United States, including those who had long argued that it posed serious risks to public health and safety, would be able to bolster their case by uttering one word: Fukushima. Three Mile Island had had the same impact three decades earlier, and the U.S. industry was only now recovering.

One thing was certain as the disturbing reports continued to arrive: this was a developing drama that would dominate the twenty-four-hour news cycle for days, if not weeks. As a result, some serious messaging would be required by all sides in the nuclear debate, and the NRC would be in the thick of it.

Shortly after 5:00 a.m. on March 11, just as word was reaching the United States of the crisis in Japan, Scott Burnell of the NRC's Office of Public Affairs e-mailed the headquarters operations officer at White Flint that he was headed in to work. "Always interesting to wake up and find huge news has occurred," he noted. At this point, he had no idea how huge.

The job of public relations, whether performed for a public agency or a private enterprise, entails dual missions: to provide accurate information and to deliver that information in the light most favorable to the presenter. Those missions can conflict. At the NRC, the Office of Public Affairs has to perform a particularly challenging balancing act. The agency is often caught in the crosshairs between critics who feel the agency is too lax on safety issues, too "pro-nuke," and an industry that pushes back against regulations it sees as too strict and has the political muscle in Washington to get its way.

Although regulatory conflicts with industry and its allies do arise, they usually occur under the media radar. As a result, the NRC is far more accustomed to defending its performance against critics in the public sector who claim the commission is too lax. And when those critics take aim, the NRC's public affairs staff tends to circle the wagons to protect its own.

Fukushima wasn't an accident in a backwater country; it was occurring in highly educated, science-savvy Japan, using technology and a regulatory playbook largely borrowed from the Americans. Thirty-one aging carbon copies of the reactors at Fukushima Daiichi were operating around the United States. Assurances about reactor safety, repeated so often in the United States, had also become the mantra in Japan. Regulators and industry were cozy—another complaint heard in both countries. Those connections were pretty hard to ignore, and the news media latched on to them almost immediately.

By 10:30 a.m. on March 11, the media calls and e-mails were pouring in to the NRC, and the Office of Public Affairs had already assembled its first set of talking points, a script of sorts to make sure everyone delivered the same message. A list of likely questions and their answers was prepared for

Chairman Jaczko, who would soon be in front of cameras fielding queries from politicians and the media. The list included the obvious:

Q: "Can this happen here?"

A: "The events that have occurred in Japan are the result of a combination of highly unlikely natural disasters. It is extremely unlikely that a similar event could occur in the United States."

Q: "Is there a danger of radiation making it to the United States?"

A: "Given the thousands of miles between the two countries, Hawaii, Alaska, the U.S. Territories, and the U.S. West Coast are not expected to experience any harmful levels of radioactivity."

Q: "Has this incident changed the NRC perception about earthquake risk?"

A: "There has been no change in the NRC's perception of earthquake hazard (i.e., ground shaking levels) for U.S. nuclear plants. As is prudent, the NRC will certainly be looking closely at this incident and the effects on the Japanese nuclear power plant in the future to see if any changes are necessary to NRC regulations."

Those were the responses marked "public answer" on the briefing materials, with this note attached: "Talk from but do not distribute." They were not to be shared without "explicit" permission from Jaczko's office. As for the NRC staff itself, the communications lid was clamped tightly; all comments had to come through the Office of Public Affairs.

Most of the Fukushima responses drafted for the chairman also contained what was marked as "additional, technical non-public information," which tended to paint a somewhat different picture of the situation. For example:

Q: "What happens when/if a plant 'melts down'?"

A: Public answer: "In short, nuclear power plants in the United States are designed to be safe. To prevent the release of radioactive material, there are multiple barriers between the radioactive material and the environment, including the fuel cladding, the heavy steel reactor vessel itself, and the containment building, usually a heavily reinforced structure of concrete and steel several feet thick." Nonpublic addendum: "The melted core may melt through the bottom of the vessel and flow onto the concrete containment floor. The core may melt through the containment liner and release radioactive material to the environment."

Q: "Will this incident affect new reactor licensing?"

A: Public answer: "It is not appropriate to hypothesize on such a future scenario at this point." Nonpublic addendum: "This event could potentially call into question the NRC's seismic requirements, which could require the staff to re-evaluate the staff's approval of the AP1000 and ESBWR [new reactor] design and certifications."

As to the question about the risk to U.S. residents from fallout, the public answer—the one that downplayed dangers because of distance—also included a non-public detail that didn't sound quite so optimistic. "NRC is working with DHS [the Department of Homeland Security], EPA and other federal partners to ensure monitoring equipment for confirmatory readings is properly positioned, based on meteorological and other relevant information."

It's unclear for whom the "non-public" information was intended, but the public responses were utilized repeatedly in coming days. When Jaczko testified before the House Energy and Commerce Committee on March 17, he offered a nearly verbatim reassurance on risks to the United States: "Given the thousands of miles between Japan and the United States, Hawaii, Alaska, the U.S. territories, and the West Coast, we are not expected to experience any harmful levels of radioactivity."

By Sunday evening, March 13, the media onslaught over Fukushima was only building. "This is a marathon, not a 50-yard dash," wrote Eliot Brenner of the Office of Public Affairs to his staff about the flood of inquiries. Brenner reminded his troops not to stray from the information contained in press releases or official blog posts. "While we know more than what these say, we're sticking to this story for now."

By Monday, March 14, U.S. reporters were beginning to home in on a logical local angle to the events in Japan: the vulnerability of U.S. reactors to earthquakes. Their instincts were good.

Six years before, the NRC had begun a review of new data from the U.S. Geological Survey about seismic activity in portions of the country once deemed at low risk of damaging earthquakes, namely the eastern and central United States. Just as in Japan, advances in seismology now were raising questions about earlier risk assessments—assessments used to site, design, and construct America's reactors. The study, conducted in partnership with the nuclear industry, was known by the shorthand name of Generic Issue 199, or GI-199.

"Recent data and models indicate that estimates of the potential for earthquake hazards for some nuclear power plants in the Central and Eastern United States may be larger than previous estimates," the NRC said in a 2010 document describing the study. And that "could reduce available safety margins" at operating reactors.

Reactors are designed to shut down automatically when a certain level of ground motion is detected, just as the reactors did at Fukushima. (Those levels are set individually for each plant, based on historical earthquake data, which in the United States—unlike Japan—dates back only a century or two.) Other systems needed to maintain the reactor in a safe state, such as coolant pumps, have to be able to work after a so-called safe shutdown earthquake. But what if that ground motion is much stronger than designers had estimated? Would all the required systems still work? Or could vital equipment fail and core cooling be lost? "Updated estimates of seismic hazard values at some of the sites could potentially exceed the design basis" for the plants, the NRC study found.

The GI-199 findings didn't gain much media attention until the magnitude 9.0 quake hit northeastern Japan. For journalists, the findings represented a Fukushima follow-up that readers and viewers in the United States could identify with. Some at the NRC weren't happy with the newfound media interest. "Frankly, it is not a good story for us," wrote Annie Kammerer, a senior seismologist and earthquake engineer at the NRC, shortly after midnight on March 15.

The day before, Bill Dedman, a reporter for NBC News and msnbc.com, had e-mailed the NRC about the seismic safety data. Dedman, a veteran investigative reporter, had trolled the massive public online archives maintained by the NRC and come across the GI-199 document titled "Implications of Updated Probabilistic Seismic Hazard Estimates in Central and Eastern United States on Existing Plants." Appendix D of the report caught his eye. The appendix contained revised assessments of the risk of earthquake-induced core damage at ninety-six reactors in the eastern and central United States. Now he had some questions for the NRC.

Hearing nothing back from his initial e-mail inquiry, Dedman sought out some math professors to help ensure that he understood the complex calculations. Early on March 15 he bypassed the Office of Public Affairs and e-mailed the authors of the NRC report, outlining how he interpreted the data. "I'd like to make sure that I accurately place in layman's terms the seismic hazard estimates," he explained. Dedman also requested information about western reactors not on the list, which the NRC provided to him that afternoon.

Early the next morning, Dedman's article, "What are the odds? US nuke plants ranked by quake risk," was posted on the msnbc.com website. Based on the NRC's estimates, the reactor at highest risk of core damage from an earthquake was not Diablo Canyon or San Onofre, located in earthquake-prone California, but rather Indian Point Unit 3, sitting thirty-five miles north of Manhattan.

The NRC's immediate reaction was to discredit Dedman's story, even while the commission's experts were poring over it. Scott Burnell e-mailed Dedman: "I understand you're making a honest effort to convey the latest research, but I have no doubt the technical staff are going to have significant problems with how you've presented it." Burnell then e-mailed Kammerer: "Apart from 'you're totally off-base,' what specific technical corrections can we ask for?"

As other media calls and inquiries about Fukushima continued to pile up in the Office of Public Affairs, the morning of March 16 was spent attempting to impugn Dedman's report, which Burnell characterized as "jaw-flapping." "Folks," he wrote at midmorning to the technical staff assigned to scrutinize Dedman's analysis, "the expected calls [about the story] are coming in—We need a better response ASAP!" The Nuclear Energy Institute was also urging the NRC to reply.

The story created an immediate stir, triggering interest from the Associated Press, CNN, the *New York Times*, New York's congressional delegation, the Congressional Research Service, and reporters around the country. (Soon after Dedman's report appeared, New York governor Andrew Cuomo ordered a safety review of the Indian Point plant, in part based on the new risk data.)

About 12:30 p.m. on March 16, Benjamin Beasley of the NRC's Office of Nuclear Regulatory Research e-mailed Burnell that the experts had come up empty-handed. "I have received no concerns or corrections regarding the msnbc.com article," wrote Beasley.

What the NRC's public affairs staff most objected to was that Dedman went a step beyond the information they had provided: he gave it meaning for a general audience. By sorting through the commission's voluminous data, Dedman had arrived at risk rankings for the nation's 104 operating reactors. (The NRC had provided the raw data for each reactor, listing the plants alphabetically but not taking the final step of ranking them in order of risk, consistent with its long-standing practice of trying to avoid identifying the most dangerous plants.) That, Dedman said in response to complaints from the NRC's Eliot Brenner, is like the U.S. Census Bureau publishing poverty figures for metropolitan areas but not then identifying the poorest of the poor "lest anyone feel bad." It's left to reporters to inform the public of the numbers' meaning. "That's our job," Dedman told Brenner.

In a conversation long after the kerfuffle, Dedman blamed the NRC's Office of Public Affairs for caring more about the agency's image than about informing the public. By attempting to discredit the messenger, versus acknowledging the public value of the information, "they lost credibility for their organization."

As the number of Japanese leaving their homes continued to climb, hastily opened shelters beyond the evacuation zones began to fill. These nuclear nomads, who fled with few possessions, were now crowded into makeshift quarters that often lacked adequate heat, food, water, or sanitary facilities. For them, the impact on daily life was immediate and burdensome.

Many had no idea why their lives had been turned on end. Without power, the ability to obtain news was gone.[2] While the rest of the world was watching the nuclear crisis unfold, those most threatened by the deteriorating conditions inside the reactors were living in an information bubble.

As the displaced populace waited for explanations, the evacuation zones kept expanding outward. In those first tumultuous days, some evacuees were forced to relocate six times or more, fleeing ever farther from the reactors, their few belongings stuffed in bags or tucked under their arms. Others, however, were forgotten, with fatal consequences.

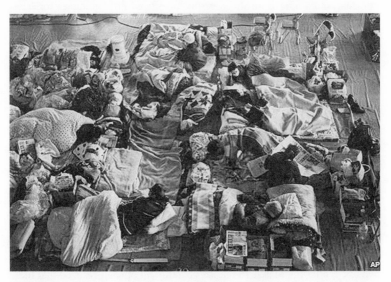

Evacuees crowded into school gymnasiums and other public buildings, where living conditions were harsh and privacy nonexistent. Many of those who fled their homes had little or no idea why they had been ordered to relocate. As conditions at the reactors worsened, large numbers of evacuees were forced to move several times. *Voice of America*

Hours after the second evacuation notice was issued early in the morning of March 12, preparations got under way to move the two hundred and nine ambulatory patients and staff out of Futaba Hospital, located about three miles (five kilometers) from the plant. Left behind, however, were one hundred and thirty bedridden hospital patients and ninety-eight residents of a nearby nursing home. The SDF reportedly were en route to transport them. Owing to a series of bureaucratic errors and communication mix-ups, the troops didn't arrive for two days, during which time the facilities had no power or heat and caregivers had departed. By then, four patients were dead. When the troops finally showed up, the patients began a grueling odyssey, spending hours on the road before the troops found a shelter that would accept them. Fourteen more died during the trip. But thirty-five patients were accidentally left behind, forgotten and not rescued until March 16. By the end of that month, officials reported that among the Futaba evacuees a total of forty patients and ten nursing home residents had died.

In a society accustomed to order and predictability, the accident response increasingly seemed chaotic and leaderless. The confusing and incomplete information coming from the government offered little guidance for Japanese seeking to understand the threat from Fukushima Daiichi. Later, officials would defend their withholding of facts by claiming they did not want to alarm people. But for many, this show of paternalism was tantamount to putting lives at risk.

Events in the scenic mountain town of Iitate, like the bungled evacuation of Futaba Hospital, came to symbolize the breakdown of the government's response and the consequences for those left to fend for themselves.

Following the explosion at Unit 3 on March 15, the prevailing winds shifted, carrying radiation to the northwest toward villages such as Iitate, population six thousand, located about twenty-five miles (forty kilometers) from Fukushima Daiichi and thus well outside the official evacuation zone. An evening snowfall blanketed the region, carrying with it particles of radioactive iodine, tellurium, and cesium. SPEEDI, the sophisticated computer tracking system, had predicted that Iitate and the town of Namie were in the path of the plume, but top officials in Tokyo dismissed the data as unreliable. Nearly two weeks after the March 15 release, government officials realized that Iitate and its neighbors had become "hot spots," with areas of radiation far above background levels. As officials debated whether evacuation was warranted, the residents of Iitate stayed put, caring for their families, their farms, and their prized cattle.

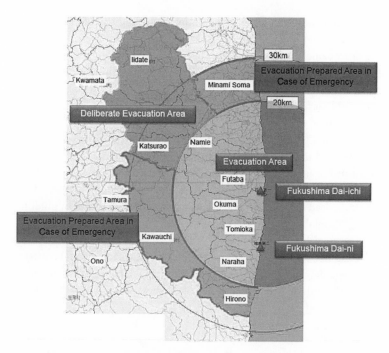

A map of the emergency response zones near Fukushima Daiichi established by the Japanese government as of the end of April 2011. The region marked "deliberate evacuation area," extending northwest of the plant to Iitate, was created nearly six weeks after the accident began. Authorities had been forced to confront the fact that winds and precipitation had spread dangerous levels of radioactive contamination well beyond the twelve-mile (twenty-kilometer) evacuation zone established soon after the accident. *Government of Japan*

All the while, radiation data from a wide radius around Fukushima Daiichi was appearing online, showing elevated levels in Iitate and other areas to the northwest of the reactors. Based on that information, some local residents grew alarmed about the lack of protective measures, expressing their frustrations to anyone who would listen. "We have been sacrificed so that Tokyo can enjoy bright lights," a tobacco farmer told a *New York Times* reporter, visiting Iitate the same day as an IAEA inspection team, dressed in protective clothing, arrived to take radiation readings among farmers working the fields. Even after the discovery of the hot spots, the government failed to act. Later, one Iitate resident expressed what many were thinking about their government: "Do they really value our lives?"

Japanese authorities, however, continued to drag their feet. Finally, on April 22, Tokyo ordered the evacuation of Iitate by the end of May. When the order came at last, the residents of Iitate were confronted with decisions both pragmatic and poignant. Some residents, especially those with young children, had already fled. A priest told a visitor, "Anyone who thinks about the future has left our village."

For others, severing ties was a wrenching experience. They already had been exposed to the radiation; had the damage already been done? Many lived on land that had been in their families for generations; could this soil be farmed again in their lifetimes? Their children's lifetimes? And what of their beloved cattle, considered almost like pets? Photos of abandoned livestock, dead or dying, in evacuation zones elsewhere had been published around the globe.

Knowing that no one would buy their animals for fear of contamination, and not wanting them to suffer, Iitate's farmers saw no choice. Before departing, they slaughtered nearly three thousand cattle. Soon, Iitate, which the year before had been declared one of the most beautiful villages in Japan, became a ghost town.

6

MARCH 19 THROUGH 20, 2011:
"GIVE ME THE WORST CASE"

On March 19, eight days had passed since the massive tsunami surged up and swamped the Fukushima Daiichi site. Ironically, what the crippled nuclear plant needed most now was water. Tons of it.

Inside the cores of reactor Units 1, 2, and 3, water levels had dropped—the result of heat boiling away coolant and of intentional releases of steam through safety relief valves to lower pressure so that makeup water could be injected. Even though seawater was being steadily pumped into all three reactor vessels, instrument readings showed that the intensely radioactive fuel in all of them remained exposed; in Unit 3 by as much as six and a half feet, by slightly less in the other units.[1] The amount of water in the spent fuel pools, especially in Unit 4, was uncertain (and a subject of continuing debate). One thing was obvious to everyone, however: water provided at a sufficiently high rate offered the best hope that operators might be able to lower the temperature within each core below its boiling point, prevent further radioactive releases, and wrestle the crippled machinery into some sort of eventual shutdown. At the least, it would buy them some time. For the past two days, Self-Defense Forces and riot police had been spraying water on and, they hoped, into the Unit 3 spent fuel pool using helicopters, high-volume water cannons, and fire engines, each capable of delivering six tons of water per hour. These were augmented by a high-pressure pumper provided by the U.S. military. As water hit the hot Unit 3 drywall head, a large steam plume shot skyward through the gaping roof, an event captured by photographers. When the NRC's team at White Flint saw the images, they were convinced that the pool, along with the Unit 4 fuel pool, was dry. TEPCO vehemently disagreed, however, and had not yet attempted to inject water into Unit 4.

Also murky was the status of the Units 1 and 2 spent fuel pools, which like the others were perched high above the containment vessels. The concrete-and-steel roof atop the Unit 1 reactor building had collapsed into the pool during the explosion there, possibly preventing water from reaching the pool, or at least filling it with debris. But the heat load from the fuel in the Unit 1 pool was smaller than the others, so it appeared to pose a less immediate threat. Unit 2, which had a higher heat load than Unit 3, still had a roof, so it was impossible to see inside. However, just doing the math—pool volume, heat of spent fuel times days without power to provide circulation and cooling—made it clear that the Units 1 and 2 spent fuel pools would soon need cooling restored or they would also be endangered.

Japan's Self-Defense Forces and riot police employed fire trucks and water cannons in an attempt to cool Unit 3's spent fuel pool. Even after the top of the reactor building had been blown off, it was unclear how much water had reached the pool. *Tokyo Electric Power Company*

Adding water to the pools was also critical to reducing radiation levels around the plant site. The water would provide a vital shield against radiation from the spent fuel; however, workers still would face a threat from the radioactive debris thrown to the ground between Units 3 and 4 by the hydrogen explosions. That and the "shine" beaming from the open pools rendered much of the area off limits. A worker on a fire truck driven close to the reactors had picked up ten rem of radiation in just two minutes—double the maximum exposure that workers in U.S. nuclear plants are allowed to receive during an entire year.

A truck known as a *kirin* or giraffe, normally used to pump concrete at high-rise construction sites, delivers water to the Unit 4 spent fuel pool. This truck was soon joined by others.
Tokyo Electric Power Company

The ability to deliver water to the pools did not improve until March 22,
when a giant truck with a fifty-eight-meter articulated boom arrived. It could
precisely direct up to 120 cubic meters of seawater per hour (about 32,000
gallons) high into the fuel pools, a far more effective aid than the aerial drops

from helicopters or the fire hoses from ground level. The truck could be driven in, parked, and then operated remotely, meaning human radiation exposure could be kept to a minimum. The single truck now in position at Fukushima Daiichi would eventually be joined by trucks from China, Germany, and the United States.

Although getting water to the reactors and pools was a high priority for the Japanese, they were also devoting considerable time and effort to restoring power to Fukushima Daiichi. Workers were slowly bringing in new lines from the outside. They finally restored off-site electricity to Unit 2 on March 20, and gradually brought power to the remaining units over the next week. However, the NRC crew had its doubts about the Japanese push to regain electricity. Connecting cables was one thing, but getting electricity flowing to equipment that had been flooded was quite another.

"We don't see where the power thing is any solution at all," Chuck Casto told his colleagues from Tokyo. In all likelihood, water damage had rendered some of the circuitry and instrumentation inside the reactors and control rooms unusable. (As it turned out, they were partly right. Even after the Unit 2 power panel was energized, it took another six days before crews could even turn on the lights in the main control room. But the availability of off-site power eventually enabled workers to switch from fire engines to electrically powered temporary pumps for delivering water to the reactor cores.)

A week after the accident, radiation levels were still so high that sending in workers to make the necessary connections and repairs to the installed equipment would be difficult if not impossible. No one had yet even been able to get close enough to the damaged buildings to map the areas of highest radiation, vital information if human workers were to be dispatched to the site. Readings from the fire truck indicated those radiation levels were deadly, however.

On the phone Casto sounded exhausted—and exasperated. His colleagues at White Flint knew he was running full-out. "Right now he's basically a 24/7 individual," Jim Wiggins explained in a briefing. "He's getting very little sleep, and he's holding up, but not for long." Casto's days—and nights—were filled with meetings, and his to-do list never seemed to get shorter. The U.S. Embassy staff was struggling to devise a contingency plan in case radiation levels rose enough to threaten Tokyo and the embassy itself had to be evacuated. Casto was being pressured for accident scenarios and dose calculations he didn't have.

Nor did the NRC at White Flint, for Fukushima was exposing huge gaps in the agency's ability to provide useful advice in real time for protecting people during a nuclear accident. The decades of stylized accident computer simulations were proving of little help in interpreting the events that had already happened in Japan, much less making credible predictions. As a result, everyone was scrambling to come up with high-confidence assessments of how bad the accident was and how much worse it could get.

Marty Virgilio, who was manning the overnight shift at White Flint, assured Casto that the NRC's Protective Measures Team was working on the dose estimates. "We're developing calculations to see what we think would be the worst case and the best case with respect to radiation levels in your neighborhood," Virgilio explained. "That would be fantastic," replied Casto. The team was aware that the French Embassy, concerned about the levels of radioactive iodine reported in Tokyo, had recommended that French nationals evacuate the Tokyo region. Ambassador Roos might well be wondering if the French knew something that he didn't.

Relations between the Japanese government and the Americans sent to Tokyo to assist with the accident response were occasionally distant and strained. This clearly frustrated the NRC crew at the embassy as well as back at headquarters. "[T]he political dynamic and the . . . organizational dynamic, is just, you know, unfathomable," said Casto over the phone. He chafed at the formalities of some meetings—"it's probably one of those pretty-face things again," he said of one coming up. These formalities, to his mind at least, ate up valuable time—and the Japanese didn't have any to spare.

As for TEPCO, dialogue between the utility and the NRC team remained practically nonexistent. And when the NRC experts finally did get the opportunity to sit down with company representatives, the talks weren't especially productive. "[T]hey said they didn't need any help and everything's in full control, under full control," John Monninger said.

While the Japanese had never anticipated an accident as complex as the one unfolding at Fukushima, neither had the NRC. This didn't stop the agency from engaging in backseat driving, though. Monninger took issue with what he saw as the utility's narrow focus in dealing with an unfolding, multipronged crisis that was far from under control. "They have one priority: Unit 3," he told White Flint. "And once they get done with that, they'll determine the next priority."

For the U.S. team, there was no shortage of priorities—and the list seemed to expand with each phone conversation.

Soon, the most contentious issue to confront the NRC would surface as the crisis worsened. High-level officials in Washington wanted to know one thing: how bad could this get? For the NRC, that was a question fraught with all sorts of implications. The agency had spent years downplaying the risks of nuclear accidents, contending that a real "worst case" could never happen. Now, it was being asked to assume the opposite.

The list of federal agencies working on the response to Fukushima in the United States and Japan included familiar names like the DOE, the Department of State, and the EPA. But it also included a small cadre of government entities few Americans have ever heard of, including the National Atmospheric Release Advisory Center. The center, known as NARAC, uses computer models and geographic data to map the spread of hazardous materials that get released into the atmosphere. Its predictions are intended to help decision makers in an emergency, and its models were capable of predicting the movement of radiation plumes with more accuracy and to considerably further distances than the NRC's RASCAL code.

At the moment, experts across the U.S. government were grappling with these questions: just what could be the worst-case scenario, and how much radiation might escape from Fukushima Daiichi if that scenario occurred? The answer had ramifications not only for Japan and the American citizens there but potentially for the United States itself. To make those predictions, the experts first had to agree on what's known as a *source term*, an assessment of how much radioactive material could actually be released from the reactors and fuel pools into the environment. The types of material and the timing of the releases were also key inputs for predicting where the radioactive plumes would travel and the nature of the damage they could cause. The bottom line issue for the U.S. government was this: what threats would such a release pose to Americans, whether at home or abroad, and what measures had to be taken to protect them? Without a more precise understanding of the source term, the answer to that question would remain frustratingly elusive.

SOURCE TERM EXPLAINED

Source term defines the types and amounts of radioactive material released during a nuclear plant accident. The source term depends on many variables, including the initial amount of radioactive material in the nuclear fuel, how much of that radioactivity gets released from damaged fuel, how much of that radioactivity is retained within the plant, and how much is released to the environment, where it can be transported to downwind communities.

Highly radioactive materials like iodine-131 and cesium-137, known as *fission products*, are by-products of the nuclear chain reaction that drives the nuclear engine. An operating nuclear reactor core contains a mixture of dozens of different radioactive isotopes. The quantities of these isotopes depend on the power level of the reactor and the length of time that the reactor has operated, among other factors.

Radioactive materials are unstable and release radiation seeking to reach a stable form. The time it takes for one-half of a given quantity of radioactive material to decay—called the *half-life*—varies from a mere fraction of a second to millions of years. In a reactor accident, the longer the onset of reactor core damage is delayed after shutdown, the more time is available for radioactive materials with short half-lives to decay into stable materials, thereby reducing the source term.

Different isotopes have different radioactive properties, which determine the relative hazards they pose to the environment and to humans and other organisms. Certain isotopes, such as plutonium-239, emit alpha particles, which cannot penetrate skin but are particularly hazardous inside the body; thus alpha emitters are very dangerous if inhaled or ingested. Others, like strontium-90, emit beta particles, which are somewhat more penetrating than alpha particles but still do more of their damage if emitted within the body. In contrast, high-energy gamma rays penetrate deeply, and hence gamma emitters like cesium-137 can do their damage from outside the body.

Different radioactive isotopes also have different chemical forms, which determine how they behave within the reactor, in the environment, and in living things. For example, beta-emitting iodine-131 becomes a gas at ambient temperature, can be transported large distances in the environment, and if inhaled or ingested concentrates in the thyroid gland, where it can deliver a high dose. Luckily, though, with a relatively short half-life of eight days, it does not persist in the environment. In contrast, plutonium-239 remains solid up to very high temperatures and is not easy to disperse, but with a 24,000-year half-life and a tendency to deposit in liver and bone, it is very persistent in the environment or in the human body.

These properties determine the relative hazards of the different stages of a radiation release. In the early stages of the accident, people can be immersed in and inhale airborne plumes. In the later stages of the accident, after the plume has passed, people can be exposed to contamination on the ground and other surfaces and in food and water. Contaminated dust particles can also be carried upward by air currents and inhaled.

When an accident causes the fuel rod claddings to overheat and break but is stopped before the fuel pellets are damaged, the release is limited to the radioactive gases trapped in the space between the pellets and cladding, including noble gases, iodine-131, and cesium-137. If overheating is not stopped in time, the fuel pellets themselves can break apart, allowing some of the radioactive materials trapped within the pellets to be released along with the radioactive gases. Ultimately, the pellets themselves can melt, and release an even wider range of isotopes, including plutonium-239 and americium-241. The greater the fraction of the core damaged, the greater the amount of material available for release.

A functioning reactor containment is designed to allow no more than a small fraction of the airborne radioactivity within the reactor—less than 1 percent—from being released. If containment venting is necessary, filters could be used to prevent radioactive materials from escaping into the environment. However, if the containment barrier is breached or bypassed, a far greater amount of radioactivity can escape.

Even if the containment fails, determining the actual amount that may be released over time is difficult because radioactive gases and particles released from damaged fuel may cool down and "plate out"—that is, stick to—various surfaces and be retained within the plant. But even these materials may eventually heat up again and escape.

Once the source term is developed, the weather conditions are defined and populations at risk are identified. Based on this information, the hazard from radioactive materials escaping from a nuclear plant can be estimated. But other variables also play a factor. Wind may blow the radioactive cloud toward or away from densely populated areas. A city forty miles away from a damaged plant may be at greater risk than closer cities if it starts to rain as the plume passes overhead.

Although the source term is used in estimating the risks, the actual hazard depends on questions that cannot be definitively answered during an accident or even long afterward. When was the fuel damaged? How much fuel has been damaged? To what extent has the fuel been damaged? How much radioactive material released from damaged fuel has been retained in the plant? Artful science is applied to estimate the source term based on the most likely answers to these important questions.

President Obama, in an address to the nation on March 17, offered reassurances. "We do not expect harmful levels of radiation to reach the United States, whether it's the West Coast, Hawaii, Alaska, or U.S. territories in the Pacific," he said. "That is the judgment of our Nuclear Regulatory Commission and many other experts." But behind the scenes, those experts were still debating the numbers, even as conditions inside the reactors and fuel pools were unclear.

Radiation monitors at California's Diablo Canyon and San Onofre nuclear plants had already picked up readings of iodine-131 just slightly above what the NRC described as "the minimal detectable activity level." It presumably was blowing in from Fukushima, 5,400 miles away. If these levels continued to increase, there was a chance that the president would later have to reverse himself and order countermeasures like banning milk shipments from certain areas. That could result in a major loss of confidence among the public.

Among the scientists and administrators, disparate views on the radiation threat abounded, apparently hashed out in the confidentiality of the White House Situation Room, where officials from a host of agencies gathered. Discussions also were going on at the U.S. Embassy in Tokyo. There had been internal disagreements in both places over the NRC's recommendation two days earlier for the fifty-mile evacuation for U.S. citizens in Japan. Now, based on aerial measurements between thirteen and twenty miles northwest of the reactors showing exposure rates above one rem over four days, the EPA evacuation standard, the NRC was confident it had made the right call. But the agency continued to take heat for the decision. On March 18, the Nuclear Energy Institute contacted the NRC to complain that the fifty-mile evacuation could undercut the public's faith in this country's ten-mile emergency planning zone.

Back at White Flint, Trish Holahan, director of security operations at the NRC's Office of Nuclear Security and Incident Response, and her colleagues were poring over some alarming dose estimates for U.S. territory that they had received from NARAC. Because the NRC's own RASCAL model had limited range, the agency had to rely on NARAC to go farther using its more sophisticated plume transport models. Employing source terms supplied by the NRC, NARAC's models were finding that thyroid doses to one-year-old children in Alaska could be as high as thirty-five rem—seven times the EPA dose threshold that would trigger the need for countermeasures such as potassium iodide administration.

But the NRC thought that some of NARAC's assumptions seemed a bit far-fetched. The potential radiation exposure of grazing dairy cows in Alaska in mid-March was one example. "The cows are kept indoors," Holahan told Jaczko over the phone. "Even the water supply is internal because they're not outside. So we eliminated that dose."

The assumption of grazing cattle in Alaska in wintertime was just one aspect of a larger issue that the NRC was trying to grapple with: the White House request for a "worst-case" assessment. From the perspective of the president, this approach made sense: if even under the most pessimistic assumptions there was little risk to the American public from Fukushima, then the president could continue to provide reassurances without fear that he would be later accused of underestimating the threat.

Joining the White House in pushing for the worst forecast was Admiral Mike Mullen, chairman of the Joint Chiefs of Staff. "I have been taught by my nuclear power community my whole life to plan around the worst case possibilities," he wrote to Obama's top science advisor. "This in great part had a lot to do in keeping our [Navy] plants safe."

However, the concept of a worst-case scenario was anathema to the NRC's way of thinking. For decades, NRC regulations and policies had been explicitly designed to avoid accounting for worst-case scenarios, which were believed to be so unlikely as not to merit consideration. Calculating a worst-case source term fell into that same category.

The NRC had already ventured outside its comfort zone when it made the recommendation for a fifty-mile evacuation around Fukushima, based on source-term assumptions that it judged very unlikely. Now it was asked to consider even more extreme cases.

The NRC instead preferred to focus on what it considered more realistic or "best estimate" scenarios. The NRC had just spent several years on a research project called State-of-the-Art Reactor Consequence Analyses meant to calm public fears about nuclear power by calculating "realistic" severe accident source terms. Those numbers were lower than previous estimates, and the NRC was pleased with the initial findings. But now, when a reasoned perspective was needed more than ever—at least in the eyes of the NRC—the White House appeared to be asking the NRC to throw its approach out the window.

The White House wanted the NRC to provide a source term to NARAC that assumed that 100 percent of the fuel in the cores of Units 1, 2, and 3 and in the spent fuel pools of Units 1, 2, 3, and 4 had melted, with the Units 1–3 primary containments failing completely so that the resulting radiation from all seven sources would be released into the environment.

Jaczko was frustrated by this request. He questioned whether the NRC was being asked to hypothesize an accident that was in fact impossible—one that would essentially vaporize all the cores and spent fuel and eject it all into the environment. "[O]bviously, there's a physical reality at some point that certain things just cannot happen," Jaczko said. As bad as this accident was, there were no plausible physical mechanisms that could vaporize the entire core of a light-water reactor like those at Fukushima. Even at Chernobyl, a more unstable type of reactor that experienced a runaway chain reaction and massive steam explosions, most of the core material remained within the reactor.

"[T]here's what's worst case and then there's what's possible," Jaczko told Holahan and the team at White Flint. "So I think what we should produce [is] a worst-case [scenario] that [is] actually possible." Instead, he said, the NRC was being asked to envision the nuclear equivalent of a "meteor hitting the earth at the same time as an asteroid strikes." Holahan agreed to go back and confer again with the reactor safety experts and come up with some new analyses.

Earlier that afternoon, Jaczko had spent forty-five minutes with the Japanese ambassador, making a combined condolence and business call at the embassy on Massachusetts Avenue. The NRC crew hoped the visit might facilitate communications between Washington and Tokyo. "We have a tremendous opportunity here now," Jaczko had told his team before heading to the meeting. "We have an ambassador who basically wants to be helpful and can pass information to help move a logjam if necessary."

Chuck Casto, back from a meeting with the chairman of TEPCO and the utility's chief nuclear officer, called his colleagues at White Flint at about 9:00 p.m. on March 18 to give them an update. "It was a cordial meeting," he said. In contrast to the few previous encounters, this time TEPCO executives asked for help from the Americans.

Although the NRC was still focused on finding a way to cool the spent fuel pools, TEPCO was more worried about the reactors themselves—in particular, the accumulation of salt inside them. As the fire trucks kept pumping in seawater, large amounts of salt were building up in the bottoms of the reactor vessels, potentially blocking the flow of the water and interfering with efforts to cool down the fuel. In addition, seawater is more corrosive to metal than the freshwater the reactors were built to use. Forced to use seawater at Fukushima, the Japanese wanted to consult about the problems it would create for them down the road. They were now asking for assistance.

Jaczko was listening in. "Do you think that project team should be NRC people or somebody else, like maybe INPO or something like that?" Casto agreed that both the NRC and the Institute of Nuclear Power Operations ought to get involved, along with the DOE, which has expertise in radiation doses and decontamination.

"I think this is a major change in the mission," Casto told Jaczko. "Sir, I believe that they are looking to us for solutions. . . . I think they were . . . desperate for options."

Jaczko's reference to a heightened industry role in the U.S. response to the accident was not hypothetical; the details were being finalized even as the men spoke. The expert industry group was set to gather at NRC headquarters. The weary White Flint team hoped the fresh troops might shift some of the load off the NRC's shoulders; with luck, the involvement of industry might even encourage TEPCO and Japanese officials to be more forthcoming with information and to accept more help.

Participants with special expertise, such as GE, which had designed the Fukushima Daiichi reactors, would be asked to provide guidance on getting water into the fuel pools, for example. Other industry representatives would be invited to help devise short- and long-term solutions in Japan.

It was not going to be simple. For its part, the NRC planned to work through diplomatic channels in Japan to come up with a cooperative agreement that was politically acceptable to Tokyo, and also to try to obtain better information about the conditions at Fukushima Daiichi and the radiation levels.

But better data would only improve things somewhat; equally important was a definitive plan of action from the Japanese. "[T]hey have shifted almost as frequently as the winds have changed there with respect to what their priorities are," Virgilio observed. "[I]t's not a good situation."

"I'm just trying to figure out who the power player is over here," Chuck Casto said. For the industry consortium to succeed, it had to be consorting with the real decision makers in Japan, and Casto wasn't sure who those were. "Is TEPCO the right organization, or should we be going to MOD [Ministry of Defense], or who?"

It was shortly after midnight March 19 at White Flint, and Casto had just left a meeting with Ambassador Roos. Roos knew he might soon face a crucial decision: if the situation at Fukushima Daiichi began to deteriorate, should he be ready to order a much wider evacuation of U.S. citizens and close the embassy? Like the White House, he was demanding better data on a worst-case scenario—information that would allow him to decide if it was safe to stay no matter how bad things got. The DOE had given Roos the results of an accident scenario that Casto had not reviewed. The NARAC models, using a source term known as the *super core*, were indicating that doses at the embassy could exceed one rem over four days, which would require evacuation under the EPA's guidelines. When Roos asked Casto about the numbers, Casto later told Virgilio, he felt "blindsided." Even the American experts advising the ambassador weren't coordinating their efforts.

In response to the previous White House request, the NRC—despite its reservations—was already developing what it considered a worst-case scenario involving Units 1 through 4. (It was also analyzing a more "realistic" scenario to have on reserve.) But now the ambassador apparently wanted projections for an even more extreme scenario in which Fukushima Daiichi Units 5 and 6 also melted down. For Virgilio, this demand went too far, given that Units 5 and 6 were relatively stable. He also presumed that the results would indicate the need to evacuate Tokyo. "[T]hat's why we're trying to do a worst-case that really makes sense given the conditions that we have now," he said. "I mean the team spent half the night last night trying to figure out where do we start from."

"Well, and I appreciate that work," Casto said. "I expect he'll turn to DOE and say, give me the worst case."

"You know the old adage," warned Virgilio, " 'Be careful what you ask for.' "

"I think they're trying to get something out to [embassy] employees to show that it's safe . . . no matter what happens—even if the extreme happens—it's safe to be in Tokyo," Casto said.

Finding the right answer—or at least a best guess—for the ambassador became even more urgent. The prevailing winds were about to start blowing toward Tokyo and continue in that direction for twelve hours.

But this was not an easy task. The debate among the NRC, the DOE, and the White House Office of Science and Technology Policy over whether to evaluate worst-case scenarios and in fact, what exactly *was* the worst-case scenario continued to percolate over the next several days, and was a source of considerable friction among the agencies. Although the NRC's emergency planning strategy included the ability to estimate source terms based on stylized cases, the urgent need for the NRC to produce something that accurately modeled a real-world event put the agency in a bind. Nobody ever figured they'd have to do it in real time for an accident as complicated as Fukushima.

At one point on March 20, the NRC's Jim Wiggins said, in apparent frustration, "I still won't let anyone use the word 'worst case' in the room here . . . because there's about five worst cases."

The White House and the DOE were griping about the NRC's performance as well. For instance, they found mistakes in the source terms that the NRC had been using. When the NRC modeled the Unit 4 spent fuel pool, it assumed that a fairly large amount of iodine-131 had been released. But the fuel most recently discharged into that pool had aged over three months. In that time, most of the iodine-131 should have decayed away. It turned out that the NRC had received erroneous information leading the staff to believe that the most recent batch of fuel had been discharged from the reactor only a month before the accident; thus, the NRC's model greatly overestimated the iodine release and resulting radiation doses to the thyroid that would be associated with a Unit 4 fire. But before this mistake was discovered, NARAC had used the source term to calculate trans-Pacific doses, finding some alarmingly high results—four rem (forty millisievert) thyroid doses to one-year-old children in California over a two-month period, for example—although these were still below the thresholds for protective action.

Interagency rivalries only compounded the difficulties. Early on, the NRC team at White Flint complained about interference from the DOE's experts at the national laboratories. The DOE later fired back with its own critique. A postaccident evaluation, prepared by two experts from Sandia National Laboratories, criticized the NRC, saying the agency "did not seem to engage aggressively until four or five days into the event." In addition, NRC emergency planning personnel were "very reluctant" to engage with their own colleagues on the research staff who had previously done analyses of events very similar to those unfolding at Fukushima, according to the Sandia study. Other Sandia experts also reportedly dismissed RASCAL as a "toy model" that should not have been used to study real-world events.

At 8:00 p.m. on March 19, Brian Sheron led a conference call updating other NRC staff members on the day's events. The afternoon meeting with the industry consortium had gone well, he reported, running beyond the ninety minutes originally planned. Industry people seemed receptive to working closely with TEPCO, offering suggestions or support. The DOE would use its NARAC resources to project dose rates in Tokyo based on the predicted wind change toward the city, but at this point there appeared no reason to alter the fifty-mile evacuation zone for Americans in Japan.

Less than an hour later, Chuck Casto was calling. "Here's today's crisis," he said to Virgilio and Sheron. After resisting requests for information and offers of help for days, now TEPCO was accusing the United States of dragging its feet.

Ambassador Roos, his staff, and Casto had just returned from a meeting with the utility. The TEPCO officials wanted to hear the NRC's assessment on radiation levels and the salt accumulation in the reactors—something TEPCO had asked for just a day earlier. "Well, honestly, I didn't have a wallet in my back pocket on that," Casto told his colleagues. "I said, 'Well, you asked me about it yesterday. There's a lot of information, a lot of analysis, and I believe we're working on that.'" TEPCO was insistent. "They basically said, we need this stuff immediately."

Now it was time for the NRC to step back a little, Casto told Virgilio. The Defense Department was willing to bring in whatever heavy equipment was needed. "[W]e're not working the logistics stuff," said Casto. "It's out of our lane." And the new industry consortium could work more closely with TEPCO.

As a trade-off for the NRC's ongoing technical assistance, the Japanese should be expected to hand over data, Casto said. "[W]e don't know the condition of the reactors . . . , what containment pressure is, what reactor pressure is, whether those things are even full. Nevertheless, all that's moot. The bottom line is get water. They need to get freshwater into that reactor." [2]

One way to start getting freshwater for the reactors would be to get desalination equipment up and running, but this would have to wait until power was restored to the site. The team back at White Flint promised to review research papers on the salt issue and get back to Casto and Monninger, who was also on the line from Tokyo.

"I'm glad you brought that up because let me make it clear," Monninger said. "We really need you guys to be the brain waves and give recommendations. [H]ere we can't really read stuff and come up with thoughts and recommendations and that kind of stuff. We want to be the, you know, the grease." The Tokyo group was too preoccupied by the never-ending stream of crises, and perhaps just too beat, to spend precious time scrutinizing scientific research. Earlier, Monninger had told his colleagues he yearned for "100 hours of sleep."

At 10:00 a.m. Sunday, March 20, industry representatives joined in a conference call with the NRC operations team at White Flint to brainstorm. The industry consortium, working out of the Marriott across the street from NRC headquarters, had agreed to send two of its technical people to Tokyo that evening or the next day to work with Chuck Casto, and then to embed them at TEPCO's emergency operations center. That, everyone hoped, might improve communications.

Almost from the outset of the accident, the Japanese nuclear industry had reached out to the U.S. nuclear industry for help, leaving the NRC as a bystander. On March 12, field representatives in Japan for GE-Hitachi, a U.S.-Japanese nuclear partnership, had contacted Exelon, the Chicago-based utility, and asked Exelon to run accident simulations for the reactors in its fleet that were of the same design and vintage as those at Fukushima Daiichi. (However, this information did not prove very useful—Exelon's attempt to model what was going on at Unit 1 on the simulator at its Quad Cities plant predicted a primary containment pressure that was only 3 percent of what was being reported.)

And the Nuclear Energy Institute appeared to have access to valuable information the NRC didn't have. The NEI had dose rates the commission had been seeking, for example. (The NRC's information often came from unexpected places. Details about radiation levels in spinach came from the *Wall Street Journal*. "It's amazing how people know this stuff and we can't seem to get it," marveled a member of the White Flint crew.) Maybe things would improve with industry bridging the gap.

The status of the Unit 4 spent fuel pool still worried the NRC team. Although on March 20 the Japanese had finally begun using fire engines and then pumper trucks to spray tons of water toward the Unit 4 pool, as they had been doing at Unit 3 for several days, workers were shooting the water from such a distance that "you have incredible losses," Monninger told his colleagues. (The arrival of the *kirin* trucks was still two days away.) "[T]he media [are reporting] that these fire trucks are going in and out, the helicopters are doing this, the super capacity pumping system. But then, when you actually [get] down into TEPCO and start talking to the engineers, you find out that it really isn't that effective." If it had been, radiation levels would have dropped, but, when pushed on the subject, the Japanese reported no change.

The NRC remained convinced the pool was dry—a view at odds with the Japanese belief. "We've got to be very careful with that because we got in trouble before by passing up that information," warned Monninger.

If the pool were in fact dry, the threat it posed was enormous. The intensely hot fuel could now be melting into the concrete pool floor. And sitting below that was the torus, filled with more than a million gallons of water. If the molten fuel reached the torus, it could vaporize that water almost instantly, causing a powerful steam explosion that could propel radioactive core material far and wide. (However, as with the NRC's earlier belief that the pool was empty, fears about the molten fuel also proved unfounded. A computer simulation the next day indicated the fuel would not be hot enough to melt through the pool floor.)

On another topic, Monninger related that after days of seeking an invitation, he and Jim Trapp had finally made it to the TEPCO emergency operations center. Its scale surprised them. "This place was massive," he said. "There's probably 250, 300 people in that room." In addition to the sheer size of the TEPCO operation, the visit to the utility's operations center was notable for another thing, Monninger told his colleagues: "There's huge [numbers of] protesters, cameras, cops surrounding the TEPCO facility." Official visitors were now whisked in via an underground garage.

Assuming the spent fuel in the Unit 4 pool was still covered with water, it would require seventy-two tons more water every day to cover losses from evaporation. The Japanese were aiming at least that much at it. The Unit 2 pool was targeted for twenty tons a day; Unit 3, ten tons a day; and Unit 1, five tons a day. But no one knew how much of this water was reaching the pools and how much was missing them and flowing elsewhere in the reactor buildings.

"One of the concerns is they turn the site into a swamp," said Monninger, "but the other is just the contamination and the runoff from all this water that's not going into the spent fuel pool."

WATER REFLECTIONS

Water propelled by tsunamis flooded the Fukushima Daiichi site, disabling the power supplies for nearly all of its safety equipment.

Lack of water damaged the Units 1, 2, and 3 reactor cores and threatened to damage fuel in the Units 1, 2, 3, and 4 spent fuel pools—if that had not already happened.

Lacking freshwater sources, workers tried to address those problems by injecting seawater into the reactor cores. But seawater corroded reactor parts more rapidly and left salt behind that might eventually block the cooling water flow.

Water dumped from helicopters and sprayed from fire trucks was intended to refill the pools and protect their spent fuel from damage.

But water that missed the spent fuel pools could end up draining down into the reactor buildings below and flooding their basements. Workers struggling to repower the Fukushima site might find their efforts thwarted if flooding submerged the safety components on the other ends of the re-run power lines.

Ironically, water took away most of the options available to responders and left them none without potentially catastrophic consequences.

The lack of information on the status of the spent fuel pools was just one of many factors causing angst and blocking interagency consensus on a plausible source term. While the NRC had thought that it finally gave the DOE and NARAC the worst-case source term they were seeking, NARAC was continuing to ask the NRC for more information. At 8:30 the next morning, the White House was convening a meeting to bring together the warring parties. The goal of that meeting was to reach a public policy decision based on science, but in the case of source terms, the science was anything but solid. There were just too many variables, to say nothing of biases.

Getting the source term right was far more than an academic exercise. The safety of American civilians and military personnel in Japan and potentially millions more in Pacific island territories, Alaska, Hawaii, and even the continental United States was involved. Was the threat real enough to distribute potassium iodide tablets to reduce the effects of radiation exposure to the thyroid? Underestimating the hazards could leave many people in harm's way. Overestimating them could result in the unnecessary movement of large numbers of people—which itself could result in casualties.

Now, it came down to whether NARAC should use a "realistic" worst case or a "worst" worst case. But who could say, given all the extreme events that had already occurred, what was truly "realistic" at this point?

To break the impasse, Jaczko's office wanted the NRC to send to the meeting not only a technical expert but also someone with senior-level credentials and experience to go "nose to nose" with Dr. Steven Aoki, deputy undersecretary at the DOE, home to NARAC, to press the NRC's position. Charlie Miller, a veteran department chief, was recruited for the task. A successful outcome, according to Virgilio, would be an "agreement high enough up that my folks wouldn't continue to bang their heads against the telephone back and forth with folks at our level about what assumptions are, and they would actually do some calculations for us."

This was likely easier said than done, however. As Mike Weber, the NRC's deputy executive director, told Miller, "probably what you're going to find out as each party weighs in is everybody has a different definition of worst case as their own . . . so we've got to come to the common agreement to go forward."

And impatience with the lack of a definitive answer was growing among a host of federal agencies—all with a stake in the decision. As Marty Virgilio put it, "DOD wants to know where to move their ships. EPA and others want to know what to expect on the West Coast. HHS [Health and Human Services] wants to know what kind of levels in order to make recommendations on whether or not they should actually recommend potassium iodide [tablets] at some point. And it goes sort of on and on."

Much of the give-and-take at that Monday morning meeting at the White House remains secret. But at the heart of the discussions was finding a way to make the best judgment call on radiation risks in the face of such uncertainty. Were conditions dire enough to warrant an evacuation order for U.S. personnel in Tokyo? Some of the data in White House hands indicated it might be necessary. At the other extreme, should the fifty-mile evacuation advisory be lifted, as some were urging? (Ultimately, the fifty-mile evacuation zone remained in place until October 2011, when the State Department reduced it to twelve miles.)

Even as Miller headed to the White House, the White Flint team was receiving important new information that could dramatically alter source term assumptions. The Unit 4 spent fuel pool, although heavily damaged, apparently had water, reducing the risk there somewhat. "Do we have any idea how we got it in there?" someone asked. The answer was no; nor did anyone know how much water the pool contained.

The NRC elected not to try to change the source term numbers at this point. "It took two days to negotiate this source term," said a member of the NRC crew. "I don't know if we want to spend another two days trying to negotiate another one."

Finally, an agreement was reached among the White House science advisors, the DOE, and the NRC on a scenario the NRC pointedly referred to as "the President's source term." This case assumed releases from three reactors and four spent fuel pools, but used "best estimate" simulations for the amount of radioactive material released from each source based on computer models of the accident using the NRC's own computer code, known as MELCOR— introducing what the agency believed was "realism" into the analysis.[3]

NARAC's results for the "President's source term" turned up one disturbing finding: a potential thyroid radiation dose to a one-year-old child on the West Coast of the United States of 4.5 rem, not far below the 5-rem EPA threshold for protective actions such as potassium iodide administration or interdiction of milk supplies. So the fears of some Americans that Fukushima could impact them were perhaps not as far-fetched as the U.S. government had led them to believe. However, because this result was below the threshold, it did the trick: even in the worst case, nothing needed to be done to protect the children in California.

NARAC also decided to use this source term to evaluate the potential doses in Japan as well, and that produced an alarming result. "You've got to evacuate [Tokyo] and everything else," reported the NRC's Jim Dyer.

This prompted the crew at White Flint to complain once again about the way the other agencies were continuing to engage in the source term exercise. "[W]e ought to just have realistic models, not these ultraconservative worst-case things," said Bill Borchardt.

But even the "realistic" models weren't simplifying matters. NARAC was also running the NRC's "plausible realistic" scenarios, which assumed far less containment damage than the NRC's MELCOR models and no releases from any of the spent fuel pools. According to these results, Japan's twenty-kilometer evacuation zone was not too small, but rather larger than it needed to be. Then came another surprise: the NRC soon discovered that it had made yet another big mistake. Its "plausible" source term was too low, even for the "realistic" case. Although the NRC tried to cover its tracks by coming up with a post hoc justification for the error, its bungling of the math did not help the commission's standing in the interagency debate.

Data would eventually show that the actual source term was greater than the NRC's "plausible realistic" scenarios but far less than the more extreme cases evaluated by the White House and the other agencies. Radioactive iodine concentrations on the West Coast of the United States never reached the levels predicted by the MELCOR source term. Tokyo at large was never imperiled except for a few hot spots, presumably created by unlikely but unfavorable local weather conditions.

However, the dose rate data did support an evacuation zone of about thirty to forty miles (fifty to sixty-seven kilometers) from Fukushima Daiichi, still a much larger distance than the twelve-mile (twenty-kilometer) zone initially established by the Japanese government or the ten-mile emergency planning zone in existence in the United States for reactor accidents.

For the NRC, Fukushima Daiichi redefined "realistic"—something the agency had stubbornly resisted for decades. Its reluctance to seriously consider the likelihood of a severe accident with a large radiological release, even for planning purposes, reflected the commission's propensity to view accident risks and consequences through rose-colored glasses.

Up until March 2011, for example, the NRC firmly believed that no realistic accident at a U.S. reactor could be serious enough to require more than a ten-mile emergency evacuation. The NRC had adopted that modest safety standard after an earlier reactor emergency provided the nation's first reality check: the 1979 accident at Three Mile Island.

Since then, the ten-mile zone had remained inviolate. In the NRC's mind, an accident like Three Mile Island—in which some fuel melted but the containment held, limiting the release of radioactivity—set the limit for the worst accident that needed to be rigorously prepared for at a U.S. nuclear plant. Now, as a result of Fukushima, the realism of this and other assumptions on safety would be severely tested.

7

ANOTHER MARCH, ANOTHER NATION, ANOTHER MELTDOWN

The March 2011 disaster at Fukushima Daiichi recalled another early spring meltdown more than three decades earlier. In March 1979, the Unit 2 reactor core at the Three Mile Island nuclear plant south of Harrisburg, Pennsylvania, suffered a partial meltdown as operators struggled over several days to establish control. The Three Mile Island accident proved much less serious than the crisis at Fukushima Daiichi, but the disasters shared much in common: design inadequacies, equipment failure, and human shortcomings. These led to inadequate cooling of the reactor cores with ensuing meltdowns, hydrogen explosions, releases of radioactivity to the air and water, and evacuations of more than one hundred thousand nearby residents.

There were also notable differences. Three Mile Island Unit 2 was a pressurized water reactor, unlike the boiling water reactors at Fukushima. Three Mile Island was precipitated by an "internal event" in industry parlance, in contrast to the "external" seismic and flooding events at Fukushima. And the challenges at Fukushima Daiichi were far more extreme, not only because of the greater scale of the crisis involving multiple reactors but because the operators had to cope with a sustained total loss of electrical power and the inability to obtain needed supplies because of damaged roads. The more extreme conditions led to a far worse outcome.

But the negligent regulatory and industrial practices that paved the way for both accidents were strikingly similar. The nuclear establishment worldwide had thirty-two years to learn from the mistakes of Three Mile Island and to find ways to avoid repeating them. Was the stage set for another disaster because Three Mile Island's lessons were forgotten?

The short answers are both no and yes. Many of the mistakes that contributed to Three Mile Island were identified after the accident and addressed in a series of regulatory reforms, with varying degrees of effectiveness. In the United States, control room instrumentation was improved, reactor core cooling and containment isolation systems were enhanced, operator training was intensified, and emergency preparedness drills were beefed up. A number of these reforms were adopted elsewhere in the world. But some critical factors that contributed to the Three Mile Island accident were swept under the rug by regulators both in the United States and abroad. These unlearned lessons remained unheeded three decades later when the waves bore down on Fukushima. Both accidents followed from one common and dangerous belief: that an accident like Three Mile Island, or Fukushima Daiichi, just could not happen.

The March 28, 1979, accident at Three Mile Island began when a pump in the system providing cooling water to the Unit 2 steam generators unexpectedly stopped running, for reasons never determined. That triggered a series of events that caused the reactor to shut down automatically from nearly full power. It was the thirteenth time in a year that problems in this cooling system had forced a shutdown. In the push to restart the reactor and resume generating profitable electricity, nobody had gotten to the root of the problem. This time, luck ran out: a combination of equipment malfunctions, worker miscues, and design flaws transformed warning flags into disaster.

When the accident began at 4:01 a.m., Unit 2 was just thirty-six minutes shy of its first birthday. The reactor was a Babcock & Wilcox pressurized water design, capable of generating about nine hundred megawatts of electricity. Three Mile Island's owner, Metropolitan Edison Company, was a small utility with little nuclear operating experience.

In a pressurized water reactor, the cooling water that flows through the core is maintained at a pressure high enough to keep it from boiling. To control the pressure, the reactor vessel is connected to a tank known as a *pressurizer*, which is normally about half filled with water and half with steam. The operators can heat the contents to increase steam pressure at the core, or they can add cooler water to achieve the reverse. Operators at Three Mile Island had been taught to make sure that the pressurizer never filled completely with water, a condition known as *going solid*, because then they might lose control of the reactor vessel pressure.

Seconds after Unit 2 shut down, three standby emergency pumps automatically started to restore the cooling water flow and resume the removal of heat through the steam generators. But a valve that had been closed for maintenance work two days earlier remained closed for reasons still unknown, blocking the needed water. The operators failed to notice the closed valve for eight minutes.

Absent heat removal by the new coolant, the temperature and pressure of the water inside the reactor vessel began to climb. The rising pressure caused a relief valve atop the pressurizer to open and discharge water to a collection tank in the containment building. Because the reactor had shut down, it was generating significantly less heat than usual. That, along with the open relief valve, allowed the pressure in the reactor vessel to drop below the point at which the valve was supposed to automatically close. But the valve stuck open, and cooling water kept flowing out of the vessel. Operators believed the valve had closed, however, because the indicator light on the control panel went off.

On March 28, 1979, a pump providing cooling water to the Unit 2 reactor at the Three Mile Island nuclear plant south of Harrisburg, Pennsylvania, stopped. It was the thirteenth time problems in this system had forced a shutdown of the year-old reactor. Small quantities of radiation escaped and concerns grew about a hydrogen explosion. Over the next several days tens of thousands of Pennsylvanians fled for their safety. *U.S. Nuclear Regulatory Commission*

This was not a scenario without precedent. In September 1977 at the Davis-Besse Nuclear Power Station near Toledo, Ohio, a sister plant to Three Mile Island, the relief valve had stuck open under eerily similar circumstances.

There was one notable difference: the Ohio reactor was operating then at a much lower power level, giving the operators more time to diagnose the problem and correct it. Unfortunately, information about that near miss was not shared with workers at the eight other reactors of similar design then operating, including Three Mile Island, or at five then under construction. The operators at Davis-Besse had failed to notice the stuck-open relief valve for about twenty minutes; at Three Mile Island it went unnoticed for more than two hours.

During this period, the open valve discharged tens of thousands of gallons of cooling water from the reactor vessel—more than half of what it held. Worse still, the operators were unaware of this because they had no means of directly observing the water level in the reactor vessel. Odd as it may seem, there was no simple gauge. Instead, they relied on the water level in the pressurizer, which was showing the amount to be rising. Something unexpected was happening to mislead them: the stuck-open valve had reduced pressure far enough that the water in the core could now boil and form steam bubbles. Much like what happens when the cap is removed from a bottle of soda that has been shaken, the expanding steam bubbles were causing the volume of the coolant to increase, forcing the pressurizer level upward even though the amount of water was decreasing.

Another set of standby emergency pumps had automatically started and were providing makeup water to the reactor vessel. This measure commonly occurred following a reactor shutdown as the rapid drop in power lowered the pressure and temperature of the water in the primary loop, causing its volume to decrease or "shrink." But the misleading water-level indication tricked the operators into thinking the pressurizer was in danger of overfilling and going solid. They turned the emergency pumps off and opened valves to drain even more water from the reactor vessel.

Design weaknesses further impaired the operators' response to the unfolding calamity. The control room computer dutifully printed out alarms and warnings, but the backlog of abnormal conditions grew so large that the printer fell more than two hours behind, jamming at one point and losing critical information. Within minutes of the start of the accident, one hundred alarms were sounding in the control room, adding to the operators' stress but providing little useful information.

The design also failed to anticipate the magnitude of the event unfolding at Three Mile Island. Radiation detectors had been installed throughout the plant; however, many of them were scaled for relatively low radiation levels. As the reactor core experienced damage, the dials on these detectors moved as high as they could go, unable to provide any useful data as radiation continued to climb. Detection of the rising radiation would have helped the operators to diagnose what was happening and to see if their efforts were working. Instead, the off-scale instruments merely told the operators they had a problem—hardly news by then.

While the barrage of unreliable information impaired the operators' ability to respond, an information vacuum hindered responses outside the plant. State and federal officials knew early on that there was trouble at Three Mile Island, but limited technology stymied their efforts to learn more. No computer links provided off-site officials with real-time data on plant conditions. Instead, they got strobe-light glimpses into the situation: a reactor pressure reading from twenty-five minutes ago, a core temperature value from ten minutes ago, and radiation levels from two minutes ago. It was like assembling a jigsaw puzzle using pieces from a dozen different puzzles. The dearth of reliable information prompted NRC chairman Joseph Hendrie to remark that he and Pennsylvania governor Richard Thornburgh were "like a couple of blind men staggering around making decisions" (prompting a strong rebuke from the National Federation of the Blind for reinforcing stereotypes).

Approximately two hours after the shutdown, the water level inside the reactor vessel dropped far enough to expose portions of the nuclear fuel rods. Some of the fuel overheated and began to melt. Its zirconium alloy cladding reacted with water to produce large quantities of hydrogen gas; some of the gas flowed through the stuck-open relief valve into the containment building. Molten fuel flowed like lava to the bottom of the reactor vessel, where it began burning through the six-inch-thick metal walls. Fortuitously, workers finally noticed that the relief valve had stuck open and closed another valve to stop the loss of cooling water.

But they found themselves struggling to replace the lost water and to restore forced cooling of the damaged core. The high pressure and the hydrogen bubbles now occupying the reactor vessel thwarted efforts to pump more water in. Finally, after several attempts and many hours, operators were able to depressurize the primary system enough to restart a coolant pump and refill the reactor vessel. They were too late to prevent about half the core from melting but in time to stop it from burning all the way through the bottom of the vessel and spilling onto the containment floor.

Around ten hours after the accident began, there was a pressure spike in the containment building—a hydrogen explosion had occurred. Fortunately, the spike was not strong enough to rupture the massive steel-and-concrete building, which retained most of the radioactivity released from the partially melted core. But radioactive material found other ways to get out.

The operators, fooled into thinking the system had too much water, had opened valves to drain more of it away. That water carried more and more radioactivity as the reactor core overheated and melted. As it moved toward four collection tanks, the water temperature and pressure decreased as it naturally cooled down, and radioactive gases dissolved in the water bubbled free. Vent lines connected the four tanks to two waste gas decay tanks. To keep the drain pathway open, the operators periodically discharged the contents of these two tanks to the atmosphere, and the radioactivity traveled along. In addition, some of the equipment leaked radioactivity into the auxiliary building, from which it later escaped outside.

The venting and other flow paths reportedly released 10 million curies of radioactivity into the air, nearly all in the form of the noble gases xenon-133 and krypton-85. (TEPCO currently estimates that Fukushima Daiichi released about 13.5 million curies of noble gases and about the same amount of iodine-131, along with about half a million curies of highly radioactive cesium isotopes. Compared with the noble gases, radioactive iodine and cesium are much more significant contributors to long-term health effects.)

It took about a month for the reactor core to become reasonably stable with its water temperature below 212°F, the boiling point. It took nearly a year for workers to be able to enter the highly radioactive containment building to ascertain the extent of the damage. It took more than a decade, and $973 million, to clean up the accident. Japanese companies and government agencies contributed $18 million and forty engineers to the cleanup effort. In about 150 minutes, a billion-dollar asset became a billion-dollar liability.

Well before that March day in 1979, American public opinion was deeply divided about nuclear power, given its safety concerns and cost overruns. The 1970s had seen a rise in protests in general, and nuclear energy triggered its own rallies. Although few Americans understood the technology, many knew it could be dangerous, and some objected loudly to its use. But by and large, even for those carrying "No Nukes" signs, the risks of nuclear power remained an abstract concept. Now, that would change.

At about 8:00 a.m. on March 28, the traffic reporter for radio station WKBO in Harrisburg, Pennsylvania, heard on his car's CB scanner that police and firefighters were mobilizing in Middletown, the river community that is home to the Three Mile Island plant. WKBO's news director telephoned

Three Mile Island and was connected directly to the reactor's control room. "I can't talk now, we've got a problem," the control room operator told the newsman, and referred his caller to the plant's owner, Metropolitan Edison Company, in Reading, Pennsylvania.

There, a spokesman for Met Ed, as the company was known locally, confirmed that a general emergency had been declared but dismissed it as a "red tape" type of thing required by the NRC when certain conditions existed. At 8:25 a.m. WKBO broadcast news of problems at the plant, relying on the utility's explanation.

A short time later, Met Ed issued a brief press release: "At 4:00 a.m. Wednesday, the reactor at Three Mile Island Unit 2 was automatically tripped and shut down due to a mechanical malfunction in the system. . . . The reactor is being cooled according to design by the reactor coolant system and should be cooled by the end of the day. There is no danger of [a] meltdown." Soon afterward, teletypes clattered in newsrooms around the United States with a short dispatch from the Associated Press: the Pennsylvania State Police had been advised of a general emergency at the Three Mile Island nuclear plant. There had been "no radiation leak," but Met Ed officials had asked for a state police helicopter to "carry a monitoring team."

In fact, by 8:00 a.m. it was clear to Three Mile Island's station manager that Unit 2 had suffered some fuel damage, based on radiation readings in the containment building. By 9:00 a.m., NRC headquarters in Washington had been alerted. Fifteen minutes later, the White House was notified— precisely the same moment that a Boston radio station reporter called the mayor of Harrisburg to ask what the city was doing about the nuclear emergency. "What emergency?" asked a stunned Mayor Paul Doutrich.

Flawed and disingenuous communication continued as hundreds of reporters from around the globe descended on tiny Middletown to record the unprecedented accident. From the outset, the journalists found that messages from the authorities were frequently at odds with each other or so cryptic as to be indecipherable. If nervous Pennsylvanians were looking for guidance, it wasn't coming from those in charge—at least in the first days. "The response to the emergency was dominated by an atmosphere of almost total confusion," concluded the President's Commission on the Accident at Three Mile Island, also known as the Kemeny Commission, in its report seven months later.

No one seemed to know what was happening or how to respond. That included state officials, who were charged with emergency preparedness; the NRC, which was assessing the accident from three different offices and eventually from the scene; and Met Ed, with officials issuing statements at the plant site, as well as in Reading, and from the New Jersey home of its parent company, General Public Utilities Corporation.

Very quickly, distrust colored nearly every exchange. At an 11:00 a.m. news conference the first morning, Lieutenant Governor William Scranton III told reporters Met Ed had assured him that "everything is under control." Even as he was speaking, however, Met Ed was venting steam containing radioactivity from the plant. Later that afternoon Scranton would tell reporters: "The situation is more complex than the company first led us to believe."

For the NRC, the telephone was the prime means of communication, but there was no dedicated line between regulators and the control room. Thus the commission had to deal in the first days with a frustratingly incomplete picture (something that might have felt familiar to the White Flint staffers hungry for information from Japanese officials three decades later). The NRC's public affairs staff was swamped by media calls; staff members often had no updated information to provide. As for Met Ed, the utility's public relations team had very little technical expertise, and the executives put forth to brief reporters soon lost credibility not only with the media but also with state and NRC officials.[1]

As the lack of reliable information stoked public fears, the NRC dispatched to the scene Harold Denton, director of the Office of Nuclear Reactor Regulation, who became the point man for the commission. Denton was handed the task of briefing President Jimmy Carter, the NRC commissioners and staff, and the hordes of media who assembled late each afternoon for a news briefing on the accident.

For many of Three Mile Island's neighbors—as for the frightened residents living around Fukushima Daiichi thirty-two years later—the uppermost concern was the threat of a radiation release, large or small. If Three Mile Island was venting radioactivity, how safe were they? It was all too apparent that the reason answers were difficult to come by was that the experts were asking the same questions themselves.

By Thursday morning, March 29, the information trickling out of Three Mile Island prompted officials in Harrisburg to raise the possibility of an evacuation, which would be the state's call. Less than twenty-four hours later, Governor Thornburgh recommended that children and pregnant women living within a five-mile radius of Three Mile Island evacuate and that schools be closed. Federal and state experts were divided on the need for such drastic action, but Thornburgh wasn't willing to take chances, fearing further radiation releases.

News of the Three Mile Island accident spread around the globe, and hundreds of reporters gathered to cover the story. Afternoon media briefings were conducted by the NRC's Harold Denton (lower left at microphones), who had been dispatched to the scene to provide updates to the media, the NRC, and President Jimmy Carter. *U.S. Nuclear Regulatory Commission*

It was the first unequivocal directive delivered to a population craving guidance. All told, nearly 150,000 people, regardless of age or gender, piled into cars and fled, eager to put distance between themselves and the troubled reactor.

Some have called Three Mile Island the most studied accident in U.S. history, at least up to that time. Two weeks after the accident, President Carter appointed the Kemeny Commission to investigate the accident's causes and recommend ways to prevent recurrence. The U.S. Senate conducted its own investigation. The NRC conducted several investigations. The U.S. nuclear industry held its own Three Mile Island postmortem. The various examiners generally agreed that the accident largely resulted from safety studies and reviews that focused too narrowly on nuclear plant designs and hardware and not sufficiently on the human part of the safety equation.

Some of the most damning language came from the twelve-member commission chaired by Dartmouth College president John G. Kemeny. The Kemeny Commission issued a blunt report in October 1979 after an intensive six-month investigation.

"[T]he fundamental problems are people-related problems and not equipment problems," the commission wrote. "[W]herever we looked, we found problems with the human beings who operate the plant, with the management that runs the key organization, and with the agency that is charged with assuring the safety of nuclear power plants." The commission also pointed a finger at "the failure of organizations to learn the proper lessons from previous incidents." As a result, "we are convinced," the commission wrote, "that an accident like Three Mile Island was eventually inevitable."

At the heart of the problem, the report said, was a pervasive attitude that nuclear power was already so safe that there was no need to consider extra precautions. The Kemeny Commission urged that "this attitude . . . be changed to one that says nuclear power is by its very nature potentially dangerous, and, therefore, one must continually question whether the safeguards already in place are sufficient to prevent major accidents."

The nuclear industry was uncowed by these conclusions. Instead, it trumpeted another finding from the report: "[I]n spite of serious damage to the plant, most of the radiation was contained and the actual release will have a negligible effect on the physical health of individuals." In the decades to follow, nuclear power supporters would rally behind this statement and repeat the shibboleth "Nobody died at Three Mile Island." This would become a huge stumbling block to comprehensive safety reform.

Still, the many investigations did result in some chipping around the edges. Among the reforms resulting from the Three Mile Island accident were enhanced training requirements for plant workers, changes in emergency response procedures, and improvements in control room instrumentation. Control room operators now spend about 10 percent of their time reviewing changes to plant procedures and refreshing their skills on full-scale simulators. Prior to Three Mile Island, plant operators typically were required to diagnose what had happened and why before they could invoke the proper response procedure. After Three Mile Island, the operators were allowed to take certain steps to counter a developing problem before ascertaining its cause. Control panels were reconfigured and on/off switches were placed near relevant gauges so an operator could quickly verify the effect of using them.

In addition, the NRC took new steps to collect and share information about problems occurring at nuclear plants. Over the years since Three Mile Island, the commission has issued thousands of notices to plant owners about design, maintenance, and operating problems encountered at reactors. Early on the NRC went further, creating the Office for the Analysis and Evaluation of Operational Data (AEOD) to formally review reports and spotlight emerging adverse trends. (The NRC disbanded the AEOD in the mid-1990s as a budget-cutting measure.)

Three Mile Island's lessons also led to changes in the nuclear industry's safety philosophy. In the 1970s, plant owners often applied Band-Aid fixes to equipment problems so reactors could quickly restart—even if it meant the problems would soon recur. But nuclear reactors aren't yo-yos, and cycling them on and off is neither wise nor cost-effective. Once companies acknowledged that, they paid more attention to finding and fixing problems that triggered reactor shutdowns, such as the recurring issues at Three Mile Island that had preceded the accident.

Today, U.S. nuclear plants operate on average at about a 90 percent capacity factor, meaning that they are almost always producing electricity, except when they are shut down for refueling, which occurs every eighteen to twenty-four months.

One direct response to Three Mile Island by the U.S. nuclear industry was the creation of the Institute of Nuclear Power Operations. Among other things, INPO functions as an information clearinghouse for the industry—and to some degree as a shadow regulator.

The déjà vu sequence of events that led to the Three Mile Island accident eighteen months after a similar occurrence at Davis-Besse was not unique. In the 1970s, nuclear utilities shared little information with each other. Companies were needlessly vulnerable to common problems because of a lack of real-time communication about operating glitches and equipment malfunctions. Now, INPO requires plant owners to share good and bad practices. The goal is to enable everyone to learn from a mistake or malfunction without necessarily having to experience it firsthand.

INPO also established standards of excellence and periodically evaluates each nuclear plant against those standards. But the sharing only goes so far. The INPO assessment reports are among the most closely guarded nuclear industry secrets in the United States. Not even the NRC gets a copy. The nuclear industry defends this secrecy on the grounds that the assessments can be brutally frank—benefits apparently missing from publicly available (and often unjustifiably tame) NRC assessment reports.

In 1993, the public interest group Public Citizen obtained confidential INPO safety reports for all U.S. nuclear plants and compared them with the assessments prepared by the NRC over the same period. Of 463 problems cited by INPO at fifty-six plants, only about a third showed up as matters of concern in the NRC's reports. INPO identified 185 specific plant problems the NRC reviews did not address, and in 115 cases the NRC praised plant performance that INPO had red flagged. A spokesman for the nuclear industry explained the differences: "The NRC's mission is to regulate the industry. INPO's mission is to be painfully candid . . . come into a plant and lay it bare."

Another downside of the secrecy—beyond hiding a useful yardstick for the NRC's own inspection performance—is that the public never knows to what extent nuclear utilities implement INPO's recommendations to fix problems.

The Three Mile Island accident also prompted the NRC to upgrade its requirements for preparing the public for nuclear plant emergencies. There had never before been a radiation release significant enough to warrant advising nearby residents to evacuate. Now, government officials—and people living near nuclear plants—were alerted to the issue.

In 1980, the NRC required that plant owners draw up evacuation plans for the public within ten miles of each plant. (Compare that with the NRC's recommendation that U.S. citizens within fifty miles of Fukushima be advised to leave.) It also mandated that biennial emergency exercises be conducted at each nuclear plant site. During the exercise, a plant accident is simulated and the Federal Emergency Management Agency evaluates the steps local, state, and federal officials take to protect the public from radiation. In parallel, the NRC evaluates how well plant workers respond to the simulated accident and work with off-site officials.

The biennial exercises are better than nothing, but not by much. In the simulation, winds are assumed to blow in only one direction, conveniently but unrealistically limiting the number of people in harm's way. The evacuations are only simulated, so there is no way to tell if the complicated logistics of evacuating all homes, businesses, schools, hospitals, and prisons could be successfully carried out. Instead, the exercises merely verify that officials have the right phone numbers and contractual agreements for the buses to carry evacuees and the hospitals to treat the injured and contaminated.

These exercises only provide an illusion of adequate preparation. As the Fukushima experience painfully demonstrated, rapidly moving people out of harm's way in the midst of a nuclear crisis is exceedingly difficult, yet critical.

Although the various Three Mile Island reviews converged on the need for major nuclear safety upgrades, there was no consensus on how wide-ranging the reforms should be. At the heart of the safety debate were these questions: Should the reforms address only the issues raised by the last accident? Or would that be tantamount to fighting the last war? If the next accident were triggered by a completely different event and proceeded along a different track, the failure of a too-narrow approach would be evident. Because of the NRC's regulatory focus on design-basis accidents that followed a certain script, it had never taken a comprehensive look at the universe of beyond-design-basis accidents—that is, everything else that could go wrong—or the need to protect against them.

The aversion to considering beyond-design-basis accidents—then called "Class 9 accidents"—dated back to the NRC's predecessor, the Atomic Energy Commission (AEC). One of the AEC's concerns was laid out by none other than Harold Denton, then a member of the AEC's licensing staff. During a January 5, 1973, AEC meeting on reactor siting criteria, he stated that "if Class 9 accidents are considered 'credible,' this may preclude the construction of reactors in the Northeast United States." In other words, if protection from reactor accidents was deemed to require large distances between the reactors and the public, there might be no suitable sites in the northeast.

However, this sentiment was not shared by the NRC's Advisory Committee on Reactor Safeguards. The panel of experts wrote in a letter to the NRC in December 1979: "The lessons learned from the TMI accident should be viewed in a broader perspective . . . there are other potentially important contributors to the probability of a reactor accident, and they should also receive priority attention."

Had the NRC followed that advice, the regulation and operation of the nation's reactors could have been transformed. But breaking out of its traditional focus into this new realm of oversight was not in the cards. Instead, in the face of Three Mile Island's evidence to the contrary, the NRC ultimately returned to its belief that beyond-design-basis accidents were rare enough to largely ignore, and it limited the scope of the subsequent regulatory reform primarily to fighting the last war.

The NRC had blown its chance to develop a comprehensive approach to preventing meltdowns and thus had failed to learn one of the most significant lessons from Three Mile Island: that if one type of beyond-design-basis accident could occur, so could others. Instead, a series of ad hoc half measures and voluntary industry "initiatives" would fill the vacuum, creating a regulatory patchwork with plenty of holes. The NRC would refuse to recognize the defects of this system for decades, until it was compelled to convene yet another task force to conduct a postmortem on yet another catastrophe: Fukushima.

8

MARCH 21 THROUGH DECEMBER 2011: "THE SAFETY MEASURES . . . ARE INADEQUATE"

On the evening of March 21, Chuck Casto left the Kantei, the prime minister's headquarters in Tokyo, feeling upbeat. The meeting had finally given the U.S. team what it had sought for days: access to "the middle layer"—the people within the government and TEPCO who knew what was going on and were willing to share information with the Americans.

Previous meetings with top-level officials of the utility and the government had not been fruitful. "You can't sit and interrogate those people and say, give me exactly what's going on, because they don't know," Casto explained. "You need that middle layer of people."

That evening, senior cabinet ministers, utility officials, and their staff experts gathered with U.S. representatives from a variety of agencies to discuss the situation. "Once we did get access to the middle layer then we really got our feet on the ground," Casto recalled of that evening. The first session went so well that the group agreed to meet the following night, and the evening conferences became fixtures, continuing for months.

The Japanese especially wanted U.S. help in devising a sustainable injection system to deliver water to the damaged reactors. The meeting provided an opportunity to exchange information on that and other subjects, something the Americans welcomed. "It felt good and successful," said Casto.

Earlier in the day, there had been a discovery that was also on the minds of the Japanese. It had nothing to do with getting water *into* the reactors. A monitoring survey had detected radioactive materials in the ocean about one thousand feet (330 meters) south of a discharge canal for Units 1 through 4. The canal carried heated water from the condensers to the ocean for cooling. Normally the water flowing from the canal was not radioactive, but it now contained radioactive cobalt, cesium, and iodine, a troubling sign that the plant had sprung a new leak.

On March 21, 2011, radionuclides, including long-lived cesium, were detected in water about one thousand feet (330 meters) south of a discharge canal (circled) at Fukushima Daiichi. The discovery threatened what had been a productive fishing area. It also heralded what would become a mounting problem for TEPCO: what to do with huge volumes of contaminated water. *Air Photo Service Co. Ltd., Japan*

Some of the thousands of tons of water pumped and dumped into the reactors to cool their damaged systems had picked up radionuclides and was now heading out into the sea. Further, the presence of iodine-131 and tellurium-132 indicated the water was leaking out of at least one core. Most of the radiation releases earlier in the accident had been airborne and reached the ocean surface as fallout from the prevailing winds. The detection of contaminants, including long-lived cesium, flowing directly into the sea posed a worrisome new problem.

Before the disaster, the waters off Fukushima Prefecture had supported a thriving commercial fishing industry. The tsunami had wreaked havoc on its fleets, ports, and processing facilities. The prospect that seafood taken from these waters might now be contaminated and unsafe to eat threatened to deliver another blow to the devastated region.

Even as the struggle to gain the upper hand at Fukushima Daiichi remained touch-and-go, the broader ramifications of the accident were becoming apparent to officials in Tokyo, to local communities, and to the Japanese public at large. In response, the government created the Nuclear Sufferers Life Support Team, with a daunting mandate. It included not only securing housing for the evacuees, who eventually would number nearly 160,000, but also organizing decontamination efforts, supplying evacuation centers, securing medical services and supplies, conducting environmental monitoring, and providing information. This, of course, came on top of efforts to help victims of the nonnuclear disaster—the earthquake and tsunami—that had battered the northeastern region.

For many Japanese, the accident became personal with the discovery on March 19 of radioactive iodine-131 in raw milk from Fukushima Prefecture and on spinach harvested in Ibaraki Prefecture, southwest of the nuclear plant. The spinach also contained trace amounts of cesium-137. The farms where the contamination was discovered were as far as ninety miles from the reactors: distance no longer guaranteed safety. Fears that the nation's food supply and its agricultural regions might be threatened added a new dimension to the accident and raised the stakes for the public and the government.

Although the levels of contamination exceeded safety limits, government officials hesitated to impose a ban. They instead sought to offer reassurance and suggest voluntary measures. Yukio Edano, the chief cabinet secretary, asserted at a press conference that someone who ate the spinach for one year would be exposed to the same amount of radiation as from a CAT scan—a comparison perhaps lost on the nonexpert.

Rather than issue an outright ban on milk sales, the Fukushima prefectural government requested that farmers halt shipments from dairy farms within eighteen miles of the reactors. The Ibaraki government similarly asked its farmers to halt spinach shipments. However, it soon became apparent that the central government had to act more assertively. On March 21, it finally banned shipments of milk from Fukushima and vegetables from Fukushima, Ibaraki, and two neighboring prefectures where these foods had been found to contain radioactive iodine and cesium above government limits.

"The levels are not high enough to have an effect on humans, so we ask that people remain calm," Edano told reporters.

Edano's announcement, meant to reassure, had just the opposite effect for numerous Japanese, who had been told to "stay calm" over and over since the crisis began. This latest development distilled the Fukushima Daiichi accident to a crucial question for many: will our food harm us? And that led to an even larger question: is our government capable of protecting us?

・　・　・

"Feed and bleed" is nuclear shorthand for a process in which makeup water is added to a reactor vessel when the closed-loop cooling system is malfunctioning. The makeup water absorbs heat given off by the nuclear fuel and is allowed to boil away, or "bleed," into the containment. Feed and bleed can also be used to cool spent fuel pools: in that case, the steam created would be released into the reactor building.

For days, the damaged reactors and spent fuel pools at Fukushima Daiichi had been kept on life support by a massive "feed and bleed" operation. If the reactor cores and spent fuel pools were all intact, the level of radioactivity in the coolant bleeding off would have been relatively low. But instead this water was highly radioactive—a clue that it was coming into contact with fuel that had sustained serious damage.

On March 24, three contractors laying cable in the basement of the Unit 3 turbine building received doses of between seventeen and eighteen rems (170 to 180 millisieverts) while standing in water. There had been water in the turbine building basements since the tsunami, but no one expected it to be radioactive. Now it appeared that heavily contaminated water was coming in from somewhere. High levels of iodine-131 suggested the source was the reactor core. This was very unwelcome news; it meant that the Unit 3 containment had a breach.

Two of the men who came into contact with the contaminated water suffered radiation burns of their feet and were hospitalized. The government had already upped the allowable dose rate for emergency workers from ten to twenty-five rems a year out of concern that the entire workforce would quickly exceed the permitted dose. Now the question became: would it have to be increased again to maintain an adequate emergency workforce?

The situation would soon worsen. On March 28, TEPCO discovered that an underground trench near Unit 2 had filled with radioactive water carrying a surface dose rate of one hundred rems (one thousand millisieverts) per hour, a level that could be lethal after several hours' exposure. Radioactive water had been detected near the discharge canal a week earlier, but the water in the trench had far higher radioactivity levels and therefore posed a larger threat. (The rate actually may have been higher; one hundred rems per hour was the upper limit of the measuring equipment.) If the depth of the water increased by another three-plus feet (one meter), it would spill out of the trench and possibly flow into the ocean. Japanese authorities theorized that the water had come into contact with melted fuel inside Unit 2, escaped through a breach in the containment, and then made its way to the trench near the turbine buildings. Contaminated water was also found in trenches outside both

Units 1 and 3, and on April 2 in a pit near the Unit 2 seawater pump, which was subsequently found to be leaking into the sea.

The fact that contaminated water was getting into places it shouldn't have been able to reach should have been no surprise given the stresses that the reactors' joints, seals, and pipes had undergone during the earthquake and tsunami as well as the later explosions—stresses exacerbated by repeated dousing with many tons of water. Until electricity was restored and cooling pumps made operable, however, the wholesale flooding of the reactor vessels and spent fuel pools had to continue. The fact that tons of highly radioactive water were being generated and were finding their way into the marine environment was an unavoidable consequence. Indeed, at this point, water was the only cure.

TEPCO was trying to capture the tainted water, but storage space was growing short. No one had ever contemplated dealing with this much water. TEPCO scrambled to find places to store and eventually treat the excess water even as workers continued to add more.

The growing volume of radioactive water would pose a Herculean challenge. The plant site would eventually become crowded with large gray storage tanks where trees once stood. Dealing with contaminated water on the plant grounds, as well as controlling groundwater flowing beneath the site toward the Pacific Ocean, would only grow more problematic as time went on.

Also pressing were the challenges of removing or reducing radiation scattered for miles across farmland and forests, in residential neighborhoods and schoolyards. The magnitude and inherent difficulty of that cleanup task were only now becoming apparent. The fallout from Fukushima—literally and figuratively—affected large regions. The evacuation zones now became decontamination zones, with varying levels of radiation, some minor, some unquestionably hazardous, dispersed randomly. "Hot spots" occasionally popped up in unexpected places. Even if it was removed from the ground, contamination could remain in trees and on hillsides to be carried eventually by rain showers or breezes to new locations. And radioactivity drawn in by the roots of trees and shrubs could be released to the atmosphere later during forest fires.

After days of failed attempts to stanch the leak in the Unit 2 pit—using sawdust, garbage bags filled with shredded newspaper, and polymer—the flow finally was halted on April 6 using a liquid-glass coagulating material. TEPCO reported that more than 137,000 gallons (520 cubic meters) of water had been discharged, carrying off iodine-131, cesium-134, and cesium-137. The leak convinced TEPCO that it had to alter its policy forbidding any discharge of contaminated wastewater. The highly contaminated water in the trenches and turbine buildings had to be drained, but there just was not enough room in the tanks on site to store it. The utility realized it would have to begin waste triage, discharging water with low levels of contamination to make space for more hazardous water. But, TEPCO noted, any discharge would require "an adequate explanation to convince the general public" it was necessary.

The release of 11,500 tons of water was announced by the utility and the government in media briefings on April 4. But Japanese authorities neglected to notify foreign governments, including neighboring China and South Korea, until two minutes after the discharge began, an oversight regarded by those governments as a diplomatic gaffe. (Among others concerned about the release was the U.S. Navy, whose vessels were deployed off the coast to help in disaster relief. Navy officials worried that the radiation might contaminate ocean water treated for shipboard use.)

April 6, 2011, was the opening of the school year in Japan, a day marked by formal ceremonies. In Fukushima Prefecture, the toll of the disaster was obvious; many school facilities awaited repairs, and playgrounds were filled with rubble and piles of contaminated dirt covered with sheeting. The number of incoming students had dwindled in many areas, although in some instances the children of evacuees living in temporary shelters were hastily folded into the class rosters. Familiar faces of teachers or administrators were missing, victims of the earthquake and tsunami or absent because they had lost their homes and been forced to relocate.

Local officials and faculty worked hard to encourage a sense of normalcy. Families wore their best clothing; banners and flowers filled the auditoriums; educators greeted students. Yet the ongoing disaster was not far from everyone's minds; normalcy was a long way off.

On April 10, one day shy of the first-month anniversary of the disaster, the Japanese government recommended the evacuation of certain areas twelve to eighteen miles (twenty to thirty kilometers) from the plant where the expected first-year dose rates would be greater than two rems (twenty millisieverts); twelve days later, it formally expanded the mandatory evacuation zone to include some areas even farther away, such as Iitate. The government could no longer ignore monitoring data revealing the presence of pockets of high radiation, especially to the northwest of Fukushima Daiichi. (Since March 15, people living in this zone had been advised to stay inside their homes with doors and windows sealed. Food supplies were often scarce—delivery trucks were reluctant to enter the area—and living conditions bleak.) The dose threshold for evacuation was chosen by Japanese authorities as the lower limit of a range set by the International Commission on Radiological Protection, which recommends that during emergencies, public exposure to radiation should be restricted within a band of two to ten rem per year.

The latest advisory signaled to many that the previous evacuation warnings had not gone far enough. That triggered another exodus of residents and generated even more doubt about the reliability of information being provided and the competence of government officials. The same day, about two thousand antinuclear protesters took to the streets in Tokyo, marching to TEPCO headquarters and then to the headquarters of the Ministry of Economy, Trade and Industry, which regulates nuclear power. (Within a few months, Friday night protests outside the prime minister's office became a regular event, sometimes attracting tens of thousands of people.) A common complaint among the early marchers: they were not getting the full story about the accident from authorities.

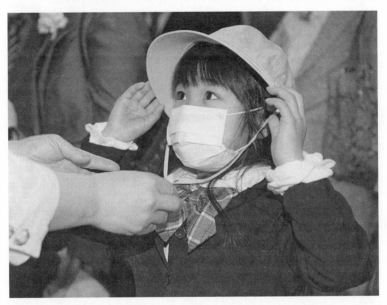

On the opening day of the school year on April 6, 2011, a first-grader puts on her hat during an elementary school enrollment ceremony in Iwaki, Fukushima Prefecture. Despite the ongoing nuclear crisis, parents and school officials worked to create a sense of normalcy for children. *AP*

Public anger in Fukushima surged on April 19 when the government authorized schools in Fukushima Prefecture to reopen provided that students attending them would not receive annual radiation doses of more than two rem. The authorities must have thought they were on solid ground with the new dose limit. After all, the government had already decided it was safe for everyone, including children, to live in areas with radiation doses of up to two rem per year. But the announcement led to puzzlement and outrage.

The government argued that students would not actually be exposed to the maximum possible radiation levels because most of their time would be spent not in the schoolyards but inside the school buildings, where they would be shielded from radiation. Not only did the public reject that explanation, but larger questions arose. If there were schoolyards in the prefecture that were so heavily contaminated they might exceed the two rem per year standard, why weren't the surrounding areas evacuated?

Another issue was why the government would allow children to bear the same radiation dose as adults. In doing so, the authorities were ignoring that children are particularly vulnerable to the harmful effects of radiation. The implications, once they sank in, set off a tempest of national protest.

A damning response came from Toshiso Kosako, a respected researcher who was serving as special advisor on radiation safety to the cabinet. On April 29, Kosako tearfully resigned at a news conference and delivered an angry statement criticizing the government for its "whack-a-mole" attitude in setting radiation safety standards. "I cannot possibly accept such a [dose] level to be applied to babies, infants and primary school students, not only from my scholarly viewpoint but also from my humanistic beliefs," Kosako wrote.

In late May, the government announced it would revert to the usual standard of one hundred millirems (one millisievert) per year. And Tokyo promised to help pay for removal of contaminated topsoil from school grounds.

In many communities, residents eager to move on with life had already launched cleanup campaigns of their own, not willing to wait for the government. Scraping away just three inches of soil could reduce radiation levels by as much as 90 percent, so volunteers dressed in hazmat suits and paper face masks used earthmoving equipment and hand shovels to strip contamination from playgrounds and other public places. They handed over to their communities thousands of pounds of radioactive dirt, covered with plastic sheeting or stuffed into garbage bags.

Clearly, water was not the only by-product of the accident that would pose disposal problems. By one official estimate, as much as 695 square miles (1,800 square kilometers) in Fukushima Prefecture—more than double the area of New York City—was contaminated with enough radiation to yield an exposure level of five hundred millirems (five millisieverts) or more per year. And the people living on or near the contaminated soil were determined that it had to be removed. They wanted to get on with their lives. As to where it would ultimately go, that was yet to be decided.

On April 17, 2011, TEPCO announced a recovery plan that it called a "road-map towards restoration." "By bringing the reactors and spent fuel pools to a stable cooling condition and mitigating the release of radioactive materials, we will make every effort to enable evacuees to return to their homes and for all citizens to be able to secure a sound life," the utility promised.

The roadmap set ambitious targets. The first, to be achieved within three months, was to reduce on-site radiation doses. The second, to be accomplished three to six months after the first target was met, involved bringing the release of radiation "under control" and "significantly" lowering it. Immediate actions were divided into three areas: cooling, mitigation, and monitoring and decontamination. The ultimate goal: cold shutdown, meaning that the temperature inside the reactors would be maintained below the boiling point, reducing the threat of pressure buildup and steam releases and providing a safety margin against future equipment problems.

Getting the reactors to a more stable state was an urgent goal. Fukushima Daiichi was still only one mechanical failure or natural catastrophe away from a second crisis. The jury-rigged feed-and-bleed cooling systems were far from robust, and safety margins were razor thin.

The recovery document served a dual purpose: it provided a technical blueprint and it conveyed to the nation that TEPCO finally had a game plan—and a timetable. Perhaps now the end was in sight.

From the perspective of the NRC's Chuck Casto, the symbolism of the plan was as important as its engineering details. "If you're in a shelter somewhere, you want to see a timeline," he said. "And from a technical point of view, it funneled everybody; [it said] this is our path. It served so many purposes to get that roadmap."

TEPCO sought the NRC's input on the recovery plan as it was being drafted, Casto said: "They listened to us on it, [we] provided advice." In his mind, the unveiling of the recovery plan was a watershed moment ranking right behind the meeting with the midlevel government and utility managers on March 21. Now, it seemed, things were moving ahead.

The timetable called for the installation of closed-loop cooling systems similar to those existing in the reactors before the accident. (TEPCO initially hoped to repair the original systems, but finally acknowledged they had suffered too much damage.) However, closed-loop systems would not eliminate the need to effectively and rapidly treat the contaminated water accumulating at the plant.

TEPCO hired Kurion, a tiny California company that employs a technology similar to the one used to treat wastewater from the Three Mile Island accident. The process uses zeolites, microporous adsorbent minerals, to bind and filter the cesium-137 in the water. TEPCO also engaged the French conglomerate Areva to develop a second phase of the treatment process, in which the residue from the Kurion system would be mixed with reagents, polymers, and sand to create a radioactive sludge-like mixture. (Some experts questioned whether the utility was merely trading one waste problem for another: large quantities of radioactive water for large quantities of radioactive glop.) Both the Kurion and Areva systems were operating by mid-June. A third system, developed by Toshiba and named SARRY, was put into operation in mid-August. The radioactive concentrates produced by these processes were stored on-site in growing rows of containers.

Although TEPCO called its plan a roadmap, it was a map that carried the utility and its government overseers through largely uncharted territory. The few precedents that existed—notably the nuclear accidents at Three Mile Island and Chernobyl—offered limited guideposts, none suggesting that the final resolution would come quickly, simply, or cheaply. Three Mile Island and Chernobyl each had involved only one reactor, so a multiplier effect was immediately tacked onto projections about Fukushima Daiichi.

The cleanup at Three Mile Island Unit 2 took fourteen years and cost about $1 billion in 1993 dollars. (Final dismantling of the damaged Pennsylvania reactor awaits the decommissioning of the still-operating Unit 1 reactor, set for some time after 2018.) And at Three Mile Island, there was no need for off-site cleanup because very little radiation spread beyond the plant property.

At Chernobyl, radiation contaminated a huge area. Although some cleanup occurred, the Soviet government exercised an option unavailable to land-starved Japan: it simply moved people out of harm's way. Ultimately about 350,000 people were resettled, and an area within a radius of about nineteen miles (thirty kilometers) around the plant remains an "exclusion zone."

Just as the Fukushima accident was unrivaled in its engineering challenges, so too was it unprecedented in its economic consequences. That TEPCO faced a huge financial toll went without saying. Two weeks after the accident, TEPCO sought a $25 billion loan from Japanese banks to help cover the cost of repairs. By mid-April, the utility was huddling with the government in an effort to devise a compensation plan for victims. The price tag ballooned. The government was estimating that the accident would cost the national economy as much as $317 billion (25 trillion yen).

On April 15, TEPCO announced it would pay "temporary compensation" of 1 million yen per household (about $12,700) to those forced to evacuate because of the reactor accident. If the evacuees needed the money anytime soon, however, they were due for disappointment. TEPCO required them to fill out three forms, one of which contained fifty-six pages and was accompanied by a 156-page instruction booklet. The evacuees, many of whom had been living in crowded shelters, were required to submit receipts for their living expenses. They were expected to provide medical records and proof of lost wages. Amid public outcries, TEPCO eventually dispatched employees to help evacuees fill out the forms, a process that required about two hours per applicant.

The growing weekly protests outside the prime minister's office in Tokyo served as a barometer of Japan's shifting attitudes toward its nuclear-dependent energy policy. But it wasn't until June 28, 2011, that the depth of that discontent could actually be gauged. The event: TEPCO's annual meeting. Well in advance, there were abundant indications that this was not going to be the normal, respectful assembly of contented shareholders in a company that the year before had posted annual revenues of nearly $54 billion.

In May, TEPCO had announced a record loss of almost $15 billion, the largest by a nonfinancial institution in Japan's history. (The loss did not include compensation claims.) President Shimizu announced he intended to resign. The utility said it would sell off more than $7 billion in assets to help cover the looming costs of compensation. On the heels of these announcements, the government made clear that it wanted more reforms. "This is just the start," said chief cabinet secretary Edano. "There must be more scrutiny and more effort."

As antinuclear protesters gathered at a nearby park, about 9,300 shareholders—the largest crowd in TEPCO's history—packed a hotel meeting room and spilled out into other rooms and hallways. Riot police helped maintain security. During the meeting, apologies from executives were drowned by shouts and jeers. The second investor to speak called on the executives to "jump into a nuclear reactor and die."

About 44 percent of TEPCO's shareholders were individual investors; financial institutions held about 30 percent of the stock and overseas investors about 17 percent. Many of the individual shareholders were elderly, including pensioners, who had seen the value of their stock plummet 90 percent since the accident.

Regardless, the renomination of all sixteen current members of the board was approved. Fifteen of them were lifetime TEPCO employees; the sixteenth—a former vice governor of the Tokyo Metropolitan Government, which was a large TEPCO shareholder—had been on the utility's payroll as a "crisis management" consultant for two years. The grueling six-hour meeting ended after a motion to shut down TEPCO's nuclear plants and halt new construction was defeated. The media reported that exiting shareholders looked exhausted and complained that their views had not been heard.

The only fresh face in the TEPCO boardroom was Toshio Nishizawa, named to replace Shimizu as president. He was hardly a newcomer, however, having spent his career at TEPCO, most recently as managing director. (Nishizawa in turn would be replaced eleven months later after a struggle over control of the company.) If TEPCO's shareholders weren't ready for a corporate overhaul, the national government was.

While the NRC team in Tokyo was doing its part to help the Japanese better understand what happened at Fukushima Daiichi—and how to move toward recovery—the staff at White Flint had been handed a mandate by Chairman Gregory Jaczko and the other four NRC commissioners.

A worried White House, members of Congress, and the American public were pressing the NRC for answers to two fundamental questions: Can an accident like Fukushima Daiichi happen here? And if so, what needs to be done to prevent it? The answers recited so often in the past—that nuclear power was inherently safe and that the existing regulations provided ample public protection—might not wash this time. After all, that's exactly what the Japanese had claimed.

The hastily formed NRC Near-Term Task Force (NTTF), consisting of six senior experts, began its review on March 30, 2011. One of the first and most obvious issues it would have to take on was whether U.S. plants were adequately prepared to deal with the kind of prolonged station blackout that Fukushima had experienced. Under the rule in existence since 1988, all American plants had to show they could cope with a simultaneous loss of off-site and on-site AC power for a certain period. At the majority of plants, the required coping time was just four hours; at one it was sixteen hours; for the remainder, it was eight.

The NRC countenanced several different approaches for coping with a blackout. One was to rely on batteries for DC power to control plant cooling systems that did not require AC power to function. Because these systems eventually would stop working even with DC power, the NRC restricted reliance on batteries for coping to no more than four hours. To prove they could cope for longer periods, plants would have to add AC power sources beyond the two emergency diesel generators they were already required to have. They could do this by purchasing additional generators or by connecting to power supplies from gas turbines, hydroelectric dams, or even adjacent reactors. (The latter option was possible because the NRC permitted licensees to assume that a station blackout would affect only one reactor at a site. Consequently, licensees could assume that equipment from a "nonaffected" reactor would be available to assist the "affected" reactor.)

Each plant determined the coping time it would need based on specific factors such as the duration of off-site power outages experienced in the past. But the NRC did not require that coping time analyses postulate extreme events that could cause prolonged outages. Nor did it require that coping strategies evaluate the possibility that alternate AC sources—like those at a reactor next door—might also become unavailable. And finally, it did not envision the possibility that flooding or fire could disable a reactor's own electrical systems, so that even if power sources were available they might not be usable.

The station blackout rule was casual about these matters in part because a blackout was considered a beyond-design-basis accident. Therefore, the requirements addressing it did not need to be as stringent.

The lax provisions of the station blackout rule were consistent with the NRC's logic: more robust protection simply wasn't needed because this kind of event was so improbable. Still, after the ten-day-long blackout at Fukushima Daiichi, these coping times appeared ridiculously low, and the agency found itself having to justify why immediate action wasn't needed to extend them.

At an NRC briefing on station blackouts on April 28, Commissioner Kristine L. Svinicki asked the staff a question that might have reflected some of what she and her colleagues were hearing, especially about the four-hour limit on batteries. "Just to a layperson, when they come to you and say, 'Is it really only four hours that nuclear power plants have to cope with some sort of event of a long duration?' . . . if you were talking to a family member, what would you say to that?"

The reply of NRC staff member George Wilson had a familiar logic: "[H]ow I've answered is that we've only had one station blackout in the United States. Our diesels are very reliable, and they restored that power within fifty-five minutes. I also explain that we have redundant power supplies. So you have to have something to take out multiple sources of power. And once I explain that . . . usually they stop, or I run overboard."

It seemed the NRC's only fallback was to say yet again, in effect, "It can't happen here." The task force had its work cut out for it.

By July 12, the NTTF turned over its first findings, a dozen multipart recommendations for the NRC commissioners to consider—hyperspeed for an agency known for taking years to debate even modest rule changes, let alone employ them. The task force dedicated its report "to the people of Japan and especially to those who have responded heroically to the nuclear accident at Fukushima," and expressed its "strong desire and our goal to take the necessary steps to assure that the result of our labors will help prevent the need for a repetition of theirs."

In directing the task force, the commission had been mindful of criticism the NRC received from the industry after the Three Mile Island accident: that the resulting recommendations were too broad and should have concerned only issues specific to events at Three Mile Island. (Others who had the opposite opinion—namely, that the corrective actions ordered after Three Mile Island were too narrowly focused—apparently were not given much weight.) This time around, the commission gave its task force a very specific scope of inquiry: "areas that had a nexus to the Fukushima Daiichi accident."

Even so, given the many similarities between the U.S. and Japanese nuclear operations, that left plenty of areas ripe for scrutiny—and, many believed, for overdue and fundamental changes. High on that list, the task force noted, was the need for U.S. nuclear facilities to better prepare for station blackouts. The blackout rule should address the possibility that a major natural disaster could disrupt both on-site and off-site AC power for *extended* periods, and knock out multiple reactors simultaneously. In other words, exactly what happened at Fukushima.

The task force recommended requiring U.S. plants be able to cool fuel for at least eighty hours during a station blackout without needing any assistance, equipment, or materials from off the site. For the first eight hours, the plant's permanently installed safety systems should be able to do the job with as little operator action as possible. This would give the operators time to set up emergency equipment, such as diesel-powered pumps and portable generators, that could be used for the next seventy-two hours, until the off-site cavalry arrived. From that point on, the plant should be prepared to cool the fuel indefinitely without power from off-site or from the on-site primary diesel generators.

The task force also emphasized that the equipment needed for that first eight-hour period should be protected to some degree from flooding beyond the design basis. It proposed storing the equipment fifteen to twenty feet above the design-basis flood level or in watertight enclosures. (The task force was more concerned about flooding than about earthquakes because it believed that nuclear plant structures could withstand beyond-design-basis earthquakes better than beyond-design-basis floods. It cited evidence that structures could survive ground shaking twice as powerful as they were designed for. In contrast, it argued that flooding was a "cliff edge" phenomenon: a plant site could be flooded even if the water level only slightly exceeded the design basis. As Charlie Miller, head of the task force, observed, it doesn't take a tsunami to create a crisis at a reactor: "[R]egardless of the way that the water gets in there it's going to cause the same effect if your equipment is not protected.")

The task force believed that these requirements should be codified in a new rule. But, recognizing that creating new blackout rules could take a long time, it urged the NRC to take interim measures by issuing orders to plant licensees. This could be accomplished by upgrading protection of the plants' so-called B.5.b equipment to ensure it would escape damage in a natural disaster.

Another issue the task force flagged was the need for reliable means to vent gases from Mark I and Mark II BWR containments should that be needed in a station blackout or other severe accident. Although the Fukushima Daiichi reactors had been equipped with hardened vents capable of withstanding high gas pressure during an accident, the vents had proven extremely difficult to operate in an extended blackout. Noting that U.S. plants would likely experience similar problems under similar conditions, the task force recommended that the NRC order plants to upgrade their vent systems.

Most of the task force's other recommendations were in a similar vein, addressing specific shortcomings in protection against beyond-design-basis earthquakes and floods, requirements for emergency plans and communications, and spent fuel pool safety. But the task force also saw a need to address the bigger picture, despite its mandate to focus only on issues with a "nexus" to Fukushima.

Priority number one, according to the task force, was to clarify the commission's "patchwork of regulatory requirements," developed "piece-by-piece over the decades," for dealing with beyond-design-basis accidents. The task force pointed out that these requirements did not amount to a set of coherent guidelines. Some issues were given higher priority than others by operators and regulators; as a result, some measures to address them were mandatory while others were only voluntary. The task force recommended development of an "enhanced regulatory framework intended to establish a coherent and transparent basis for treatment of Fukushima insights." Although the report did not say it explicitly, the implication was that the current regulatory framework was incoherent and opaque.

Despite these criticisms, the task force asserted that the "continued operation and continued licensing activities [for new reactors] do not pose an imminent risk to public health and safety." If the task force had concluded otherwise, it would have set off a firestorm. Still, the mixed messages weakened the overall impact of the report as a driver for change. Elsewhere, however, the Fukushima accident prompted some soul-searching and surprising declarations.

On May 6, Prime Minister Naoto Kan asked the owner of the Hamaoka nuclear plant, located about 125 miles (two hundred kilometers) southwest of Tokyo, to shut down two reactors and to refrain from restarting a third. Hamaoka sits atop a major geologic fault and seismologists believed there was a high likelihood of a magnitude 8.0 earthquake in the area in the next thirty years. Hamaoka had long been considered the most dangerous plant site in Japan because of its location. Although the plant's owner, Chubu Electric, initially refused Kan's request, it agreed a few days later.

The government's action was seen by some as a harbinger of things to come. Would the Kan government make similar requests of other plants? Would reactors currently shut down for routine inspections be allowed to restart?

In early July, the Kan government dropped the other shoe. To reassure the public, Japan's reactors would be subjected to a two-stage safety check known as *stress tests*. The first stage would be conducted at the nearly three dozen reactors currently out of service for maintenance or other safety issues to determine their ability to withstand large earthquakes and tsunamis. The results of that stage would be used to determine whether plants could restart. The second stage would be a more comprehensive review, and its results would determine whether plants should continue to operate. (Despite their name, the stress tests were merely paper studies. No actual stressing of facilities was involved.) Utility officials were required to submit the first-stage test results to Tokyo by the end of October. Only after government approval could the reactors be returned to service. Any plans for a quick restart during the summer of 2011 were now on hold.

Kan wasn't finished. A few days later, before a national television audience, he called on his country to phase out its reliance on nuclear power.

In his view, the safety myth surrounding nuclear power was exactly that: a myth. "Japan should aim for a society that does not depend on nuclear energy," Kan declared. "When we think of the magnitude of the risks involved with nuclear power, the safety measures we previously conceived are inadequate." To all outward appearances, he had become a nuclear apostate.

Kan was echoing sentiments taking root across Japan. Nearly three quarters of the public supported an energy policy that would eliminate nuclear power altogether, according to one survey. In an editorial, the *Japan Times* noted that nuclear power "worked for a while, until, of course, it no longer worked. Now is the time to begin the arduous process of moving towards safer, renewable and efficient energy resources."

One newspaper described Kan's phaseout announcement as "a complete turnaround of the government's basic energy plan." And indeed it was. Back in 2010, the cabinet had approved a Strategic Energy Plan that called for building fourteen new reactors by 2030, which would mean that half the nation's electricity would come from nuclear power. Only China was planning a more aggressive construction program.

The nuclear push was designed to give Japan greater energy security—the rationale for initially embracing the atom decades before. A larger fleet of reactors would provide a cushion against supply or pricing problems with imported fuels, such as oil. However, nuclear power would not deliver energy independence, because the Japanese imported almost all of the uranium needed to fuel the nation's plants. To provide a secure domestic supply of nuclear fuel, Japan was intent on developing fast breeder reactors and the reprocessing plants needed to provide them with plutonium fuel. In practice, however, these facilities were proving technically challenging and extremely expensive.[1]

In his address, Kan conceded that the phaseout of nuclear power would not happen overnight. And—aside from endorsing more renewable energy sources—he failed to outline how Japan might meet its huge energy needs once the nuclear plants were shut down.

Kan's motives immediately came into question. His popularity was at a record low, the result of what was regarded as his ineffectual leadership during the accident. Was this a politician's last hurrah—an attempt to burnish a tarnished legacy? Or was it the response of a leader who had experienced firsthand the dangers inherent to nuclear power and now wanted to steer a new course? Opinions were deeply divided.

Kan had promised to step down once the accident recovery was under way. As a lame duck, he lacked the political capital to institute a nuclear phaseout, regardless of his motives, and a spokesman subsequently clarified that Kan was merely announcing his personal views. Two weeks later, however, the government agreed to his proposal and backed his plans for reducing Japan's reliance on nuclear power.

In late August 2011, Kan resigned, his fifteen months in office marked by the worst crisis Japan had faced since the end of World War II. The task of steering the nation to a new energy policy would fall to his successor, Yoshihiko Noda, a fellow member of the Democratic Party of Japan. In a speech to the Diet in mid-September, Noda outlined his energy plan. Despite public opposition, he promised to restart idled reactors by the following summer, saying it was "impossible" to sustain Japan's economy without them. As for a rapid phaseout of nuclear power, that also appeared unlikely. "It's still too early to say if we can get to that stage," Noda told the *Wall Street Journal*.

Gregory Jaczko may have hoped that quickly assessing the lessons of Fukushima and devising an appropriate response would be a straightforward task, but that view apparently was not shared by his fellow commissioners. Jaczko had hinted he might face some opposition, especially concerning the ninety-day deadline he had set for the commission to establish its priorities. During a speech at the National Press Club in Washington in mid-July, he was asked whether he had his colleagues' support for his aggressive timetable and agenda. "Well, we'll see," he replied.

Turns out he didn't. The next day, when the five NRC commissioners sat down with the task force to have their first public discussion of the report, two of them promptly expressed doubt about the need for fundamental changes in regulation. Commissioner Svinicki asked whether some of the task force's recommendations, notably those calling for increasing safety margins as a hedge against uncertainty, represented a "repudiation" of the NRC's increasing reliance on "risk-informed regulation."

Commissioner William C. Ostendorff also took exception to the need for a major overhaul. "I personally do not believe that our existing regulatory framework is broken," he said. And, he added, any policy changes needed to be done in consultation with "our stakeholders."

The largest and most influential of those stakeholders, of course, is the nuclear industry. And that industry has always believed that the best defense is a good offense. Its leaders were hurriedly organizing their Fukushima response, hoping to head off new rules. From the industry's point of view, voluntary actions it devised on its own were preferable to mandatory ones handed down by the NRC, and it soon put forward its own ideas. (This tactic was nothing new. The NRC's embrace of industry-proposed measures over many years was responsible, in part, for the patchwork of regulations criticized by the task force and others.)

The industry's answer to Fukushima was a plan it called FLEX, shorthand for "diverse and flexible mitigation capacity." FLEX envisioned a rapid-deployment force of portable equipment such as backup pumps, generators, batteries, and chargers that would be prestaged at or near nuclear facilities. The goal was to provide redundant equipment to keep reactor fuel cool for a certain period in the event of a prolonged station blackout. The industry patterned FLEX after a response to the September 11, 2001, terrorist attacks and the NRC's B.5.b order requiring emergency backup equipment in the event of a fire or explosion caused by an airplane crashing into a nuclear facility.

Although the B.5.b equipment has been touted as an added layer of safety in the event of a crisis, post-Fukushima inspections by the NRC showed that at many sites the backup equipment would be unlikely to function at all during a severe event, especially one involving a natural disaster such as a flood or an earthquake. This should not have been a surprise, as the NRC had not required that the equipment be safety-grade, or "hardened," to withstand either design-basis or beyond-design-basis events. (In other words, the B.5.b equipment could legitimately have come straight off the shelf from Home Depot. Safety-grade components, on the other hand, must meet stringent quality standards and be rigorously tested to confirm proper performance.)

Now, however, the FLEX program was being promoted by some as "B.5.b on steroids." A better description might be "B.5.b on fertility drugs." Instead of hardening the B.5.b equipment to safety-grade standards or beyond, the FLEX approach would simply add more unhardened items. Utilities would place multiple units of equipment at diverse locations on- or off-site in the hope that no matter what the catastrophe, something somewhere would survive to cool the reactor core and spent fuel pools.

Even though the FLEX approach would require the purchase of more equipment, it would save the industry money because nuclear safety-grade standards are costly and difficult to meet. As Charles Pardee of Exelon Generation Company summed it up in a December 2011 public meeting, "it's cheaper to buy three [pumps] than one and a heckuva big building [to put it in]."

While the NRC quibbled over how to deal with the task force's recommendations, the summer of 2011 was producing moments that kept nuclear safety on Americans' radar. A flood and an earthquake—smaller than the natural disasters that had struck Japan—threatened two nuclear plants. These were the kinds of events that nuclear operators viewed as so unlikely that they could be written off. Until they happen.

The Fort Calhoun Nuclear Generation Station, north of Omaha, Nebraska, during heavy flooding on the Missouri River in June 2011. A year earlier, the NRC had cited the plant and its owner, the Omaha Public Power District, for an inadequate flood protection plan. As a result, new flood barriers were installed and the plant survived the 2011 flood undamaged. *U.S. Nuclear Regulatory Commission*

In mid-June, the Missouri River, swollen by record snowmelt and heavy spring rains, sent floodwaters surging across much of the upper Midwest. The Fort Calhoun Nuclear Generating Station, about thirty miles north of Omaha, Nebraska, went from sitting alongside the Missouri to sitting in it. The NRC had issued Fort Calhoun its operating license in 1973 based on representations by the plant's owner, Omaha Public Power District, that the plant could withstand flooding up to 1,014 feet above mean sea level. Years later, however, NRC inspectors discovered that flooding above 1,008 feet could disable vital equipment in several structures.

The key to Fort Calhoun's flood safety plan was old-fashioned sandbags piled atop floodgates. NRC inspectors had previously determined that the floodgates could not support a five- or six-foot-high stack of sandbags. The utility's own risk assessment concluded that "severe core damage results if either intake or auxiliary building sandbagging fails." The possibility of flood-water penetrating the plant walls was an additional threat.

Omaha Public Power District argued that the chance of a flood exceeding 1,007 feet above mean sea level was so small that the sandbagging response was adequate as it stood. Nevertheless, the NRC cited the plant in 2010, forcing the plant to install new flood barriers. But in June 2011, the operator of an earthmover accidentally broke through a newly installed flood berm, allowing the rising Missouri to pour into the plant site. Water reached building entrances. "It was a jarring sight . . . a boat tied to the nuclear plant," said the local congressman who toured the Fort Calhoun plant, traversing catwalks to gain access. Although water levels rose around the plant entrances to a depth of two feet, and operators relied on backup generators, the plant was not damaged because of the newly upgraded flood barriers.

And then came the earthquake that surprised just about everybody. At 1:51 p.m. on August 23, 2011, a magnitude 5.8 quake rattled central Virginia, its epicenter about thirty-eight miles northwest of Richmond. The largest previous earthquake in this zone was reported in 1875 and was estimated at a magnitude of 4.8. One of magnitude 4.5 had produced minor damage in 2003.

The North Anna Power Station sat approximately twelve miles from the epicenter. Ground motion exceeded what the plant was designed to withstand—making this a beyond-design-basis accident. Although North Anna suffered no serious structural damage, it did temporarily lose its connection to the off-site power grid, just as occurred at Fukushima Daiichi. Four backup diesel generators automatically started and provided power for nearly four hours (although one sprang a coolant leak and a replacement had to be located). Power wasn't fully restored for nearly nine hours.

In both near misses, the plant owners and the NRC pointed to the lack of damage as proof that the safety margins built into U.S. reactors and regulations were adequate. That kind of logic, critics have long said, is akin to arguing that if a drunk driver makes it home safely, the public doesn't need to worry about drunk driving. At both Fort Calhoun and North Anna, the owners had taken extra steps in advance that would head off serious damage from floods or earthquakes. In the case of North Anna, the owner, Dominion, voluntarily upgraded seismic protection at its two units in the 1990s when it learned that earthquakes posed a greater threat than previously known. That made Dominion the exception among nuclear utilities. Even though North Anna had survived intact, there was no assurance that the next plant experiencing a rude surprise would be as lucky.

After the quake in Virginia, an expert offered a takeaway lesson. "[W]hat I would say in terms of lessons learned from Fukushima and now yesterday's quake [at North Anna] is that setting reactor design . . . hazard limits just above recorded human experience is turning out to be really shortsighted," said Allison Macfarlane, a geologist and environmental policy professor at George Mason University. "With something like a nuclear reactor," she told a reporter, "I would like a large safety margin."

Macfarlane's opinions would soon carry additional clout. In mid-2012, she would take over as chairman of the NRC. The threats posed by earthquakes, she promised, would move up on the NRC's priority list.

As the NRC Near-Term Task Force moved ahead with its assessment of the lessons from Fukushima, the agency itself was wrestling with growing internal dissension. Even before Fukushima, Jaczko had told friends and acquaintances that he felt isolated on the commission, believing himself to be the lone voice for tougher oversight. (Commission votes were often 4–1.) But there were also complaints about Jaczko's management style. He was considered brusque toward staff members and his fellow commissioners, and some said he was prone to intimidating those who disagreed with him.

As chairman, a position he assumed in 2009, Jaczko had authority over commission activities related to budget and administrative matters. However, any new policy decisions and safety regulations require a majority vote, and it was clear as the months went by that the other commissioners disagreed with the chairman about the extent and the urgency of regulatory reform.

In mid-December 2011, the infighting at the NRC became public at a hearing before the U.S. House Oversight and Government Reform Committee, where serious allegations levied against Jaczko were aired. Two months earlier, Jaczko's fellow commissioners—two Democrats and two Republicans—had written to White House Chief of Staff William Daley accusing Jaczko of causing "serious damage" to the agency. The letter expressed "grave concerns" about his leadership and management style. They contended that Jaczko had improperly invoked emergency powers during the Fukushima accident without consulting his colleagues, and said he had set an agenda and timetable for the NTTF review that went beyond what the majority believed necessary. The letter also claimed that he "intimidated and bullied" staff. The NRC's inspector general investigated these allegations and ultimately exonerated Jaczko of overstepping his legal authority, but found instances in which his behavior was "not supportive of an open and collaborative work environment."

The inspector general's findings were unflattering but hardly federal offenses. By the time the report was issued in June 2012, however, the political damage had been done. Jaczko had announced his resignation from the commission a month earlier. Allison Macfarlane, pledging to run the NRC in a "cooperative and collegial manner," was named by the Obama administration as his replacement.

In Japan during the summer of 2011, the magnitude of the recovery task grew ever larger, more daunting—and more costly. Although few details of the conditions at Fukushima Daiichi were made public, news was pouring out about the extent of contamination to the surrounding area and how long it would take to get back to normal.

The government was saying that up to 1,500 square miles (four thousand square kilometers) had been contaminated to the extent that it may require cleanup. (Radioactive substances from Fukushima Daiichi ultimately would be detected in all of Japan's prefectures, including Okinawa, about one thousand miles from the plant.) On August 26, the minister in charge of the crisis response announced that over the next two years the government would cut radiation in the affected area by half.

The goal, said the minister, Goshi Hosono, was to bring radiation levels below two rems (twenty millisieverts) per year, which had been the threshold for evacuation but was still twenty times higher than the previous standard for public exposure. "[W]ith enough government funding and effort, it can be done," he pledged. Hosono also promised that the government would bear the costs of the cleanup, which some experts thought could be as high as $130 billion.

The passage of time—and natural decay of radioactivity—would do most of the work; humans would speed nature along by removing soil, plants, and trees. In areas where children might be exposed, the goal was to reduce radiation by 60 percent.

For the tens of thousands of people still living in temporary shelters and for families who had spent the summer with their children cooped up indoors to minimize exposure, Hosono's announcement signaled progress. But the good news came mixed with the bad. "Some places may have to be kept off-limits to residents for a long period of time even after cleanup operations are undertaken," Hosono said at a media briefing. Comparisons to the permanent exclusion zone around Chernobyl were unavoidable.

Radiation data gathered by the Ministry of Education and Science showed pockets of extremely high readings across the contaminated zone. In the town of Okuma, two miles southwest of Fukushima Daiichi, some areas recorded radiation levels in excess of fifty rems (five hundred millisieverts) per year.

As much as the government—and many dispossessed residents—were pushing to repopulate the communities now standing empty, it was obvious that no amount of scraping or scrubbing or isotope decay was going to make certain areas safe. Decontamination efforts simply were not that effective; typically they could reduce the dose rate only by about one-third. Ironically, some locations within the twelve-mile (twenty-kilometer) evacuation zone had radiation levels above those now measured at Fukushima Daiichi itself. In certain places at the plant, radiation levels were dropping, thanks to the massive cleanup effort. The reactors, however, remained unstable and highly radioactive.

The details about contamination and the risk it continued to pose spilled out from government reports and media accounts as summer turned into fall. In August, low levels of cesium were detected in a rice sample taken ninety miles from Fukushima Daiichi. A sample of beef from Fukushima Prefecture was found to contain high levels of cesium. In Tokyo markets, food shoppers saw radiation levels marked alongside the prices of their favorite fruits and vegetables. And those eager to prey on public fears, especially among families with young children, peddled their own products, including a $6,500 bathtub that was touted as being able to soak away radiation.

It was becoming increasingly obvious that TEPCO was on the brink of financial collapse and would soon need help from the Japanese government— and taxpayers. Compensation claims alone could exceed the company's assets, predicted *The Economist*, which went on to say: "Only the government can save TEPCO from bankruptcy." The unanswered question, it said, was whether the government would impose wholesale reforms within the company in exchange for a bailout.

TEPCO was not winning any allies among the Japanese public. In early October, the utility announced a 15 percent rate hike for customers. Although the first compensation checks to evacuees began arriving at about the same time, TEPCO seemed unwilling to fully acknowledge the damage it had caused. In fact, in one instance, the utility denied "owning" the radiation causing the contamination.

The proprietors of the prestigious Sunfield Nihonmatsu Golf Club, located about thirty miles from Fukushima Daiichi, sued TEPCO, seeking damages to clean up the closed course. The utility countered with a novel defense: the radioactive substances that fell on the course "belong to the landowners and not TEPCO." "We are flabbergasted at TEPCO's argument," said a lawyer for the club. The utility also argued that radiation levels on the golf course were below allowable levels set for schoolyards, and thus were not a hazard. A lower court agreed with TEPCO. On appeal, TEPCO's denial of ownership of the radiation was rejected, but the claim for compensation was turned down on the grounds that if the radiation levels were safe for schoolchildren, they were safe for golfers.

On November 4, TEPCO got a lifeline from the government in the form of an $11.5 billion bailout. It came with strings attached. The company agreed to cut 7,400 jobs and $31 billion (2.5 trillion yen) in costs. The government asserted it would expect more, including a "thorough reorganization," Edano told reporters.

Edano, the former chief cabinet secretary, was now minister of Economy, Trade and Industry in the cabinet of Prime Minister Yoshihiko Noda. (Edano stepped in as minister after the abrupt departure of the original appointee, Yoshio Hachiro, who, one week into office, joked to reporters that communities near Fukushima Daiichi were "dead towns." He resigned.)

While Japanese officials had access to TEPCO's bleak financial picture, the government still remained largely in the dark about what actually had happened at Fukushima Daiichi. On that, TEPCO was stonewalling. Nearly nine months after the disaster, the company finally began revealing that the accident was far worse than it had previously acknowledged.

Clearly, Fukushima Daiichi had come close to a truly major catastrophe. Using computer simulations, TEPCO estimated that the fuel rods in Unit 1 had completely melted through the reactor vessel and eaten through 6.5 feet of the 8.5-foot-thick concrete floor of the containment structure. At Unit 2, more than half the fuel had melted, and at Unit 3, almost two-thirds had melted. The fuel in all three was now sitting in the bottoms of the containment structures. Ongoing pumping operations were keeping the fuel below 100°C (212°F), a threshold at which it no longer posed the threat of boiling dry. But this was not a reliable system; another earthquake could send it careening out of control again.

A few days after revealing its accident findings, the utility released its own assessment of its performance during the disaster. The key conclusion: TEPCO had made no significant errors. It was a self-serving review written by executives and a handpicked committee. TEPCO stressed its view that the tsunami, not the earthquake, was the direct cause of the disaster and said its operators had made no mistakes in dealing with the crisis. The utility also retracted an earlier statement that an explosion had breached the Unit 2 containment on March 15. This statement had contributed to confusion about the source of the large release that day that caused radiation to spike at the plant and contaminated the large area to the northwest. (The report did note that many details were still unknown.)

Other examiners took a more jaundiced view of TEPCO's role. In coming months, the accident would be scrutinized by multiple investigating committees, none of which would hold TEPCO blameless. The period ahead for the utility would be rocky.

Although the reactors seemingly had been pulled back from the brink, there were constant reminders of the precarious condition of the plant. Areas of high radiation precluded repairs or even inspections, meaning the status of equipment inside stayed a mystery. Mountains of debris, some of it badly contaminated, remained piled up around the facility. The jury-rigged cooling systems were vulnerable to shocks ranging from another earthquake to freezing winter temperatures that could cause ruptures of piping exposed to the elements.

On December 16, 2011, nine months and one week after the disaster struck, Prime Minister Noda went on national television to announce that the situation at Fukushima Daiichi was "under control." The plant had achieved cold shutdown and the reactors were stable, he said.

Some experts called the assertion premature, motivated more by political exigencies than engineering certainties. "The plant is like a black box, and we don't know what is really happening," an official of a neighboring town told the *New York Times*. "I feel no relief." Nor, apparently, did a huge crowd of protesters who took to the streets the next day in Tokyo, banging drums, waving signs, and chanting "No Nukes."

As 2011 came to a close, about 180 police officers and firefighters made one final trip along the rugged coastline of Fukushima Prefecture, searching for bodies of people missing from the earthquake and tsunami.

Although the reactor accident itself was not directly responsible for any immediate deaths, the chaos and releases of radioactivity of the first weeks had hindered rescue efforts. Might some of the tsunami victims have been saved if rescuers had gained access sooner? It was a question often asked, especially in communities near the reactors.

Now, as the dreadful year was ending, emergency workers looked among the rocks and along the breakwaters. Some wore protective clothing, for their search took them inside the evacuation zone around Fukushima Daiichi. Japan's official death toll from the tsunami was 15,870 people, with nearly 2,800 missing. More than two hundred of the missing were from Fukushima Prefecture. No more bodies were found.

9

UNREASONABLE ASSURANCES

On March 15, 2012, the five NRC commissioners sat in a row in a Senate hearing room, summoned by the Environment and Public Works Committee and its forceful chairman, Barbara Boxer of California. The hearing topic: "Lessons from Fukushima, One Year Later."

Despite the title, some committee members questioned whether the United States really had anything to learn from Fukushima. When it was his turn to take the microphone, Senator John Barrasso of Wyoming, the ranking member on the Environment and Public Works nuclear safety subcommittee, decided to play devil's advocate. The NRC's critics were saying, he noted, that unless the agency took stronger measures to address the vulnerabilities revealed by Fukushima, then "it may be only a matter of time before a similar disaster happens here." Pointedly skipping over NRC chairman Gregory Jaczko, Barrasso asked the remaining commissioners to respond to the prediction.

First to reply was Commissioner William Magwood, who declared: "I think that our infrastructure, our regulatory approach, our practices at plants, our equipment, our configuration, our design bases would prevent Fukushima from occurring under similar circumstances at a U.S. plant. I just don't think it would happen."

Continuing down the line, commissioners Kristine Svinicki, George Apostolakis, and William Ostendorff echoed Magwood's assessment.

Technically, the four commissioners were correct in asserting that "it can't happen here"—if "it" means an event *exactly* like Fukushima, involving a magnitude 9.0 earthquake, a fifty-foot tsunami, four reactors in crisis, and multiple meltdowns and explosions. For one thing, no U.S. nuclear plant site now has four operating reactors. But as for a different series of calamities triggering a core meltdown, containment breach, and widespread land contamination? The chance of that happening is low—but it isn't zero.

One year after the accident at Fukushima Daiichi, the five nuclear regulatory commissioners appeared before a Senate committee, where they were asked if a similar disaster could happen in the United States. The commissioners, from front to rear, were William Magwood, Kristine Svinicki, NRC chairman Gregory Jaczko, George Apostolakis, and William Ostendorff. *U.S. Nuclear Regulatory Commission*

For example, many reactors in the United States are downstream from large dams. A dam failure, whether caused by an earthquake, a terrorist attack, or a spontaneous breach, could rapidly flood one of these plants with little warning, compounding any problems caused by the same event that breached the dam. Although the NRC had known for many years that it was underestimating the threat of dam breaches, the agency was taking its time deciding what to do about it. To some NRC staff members, the catastrophic flooding at Fukushima was a painful reminder of this unresolved vulnerability at home.

IT *COULD* HAPPEN HERE

The terrorist attacks of 9/11 reminded the U.S. government of the threat of catastrophic sabotage against critical infrastructure targets. Like other federal agencies, the NRC began to reassess the vulnerabilities of nuclear plants. Those included more than terrorists piloting jetliners. An attack on a dam located upstream of nuclear facilities posed a hazard greater than previously thought. And the terrorist threat alerted the NRC to the dangers of accidental failures of upstream dams as well.

Citing domestic security concerns, the NRC concealed for many years its growing worry about the threat to reactors posed by a dam collapse. Thirty-four reactors at twenty sites around the country are downstream from large dams.

A dam failure could rapidly inundate a nuclear plant and disable its vital power supplies and cooling systems. The risk of such failures was not taken into account in the design-basis flooding analyses when the plants were licensed. Other causes of flooding, such as rainfall, were considered, but these pose far less risk because the water would rise more gradually, providing greater time to prepare.

The issue became public in 2012 when an NRC whistleblower accused the agency of covering up information about the vulnerability. One plant appeared especially at risk: the three-unit Oconee Nuclear Station in South Carolina.

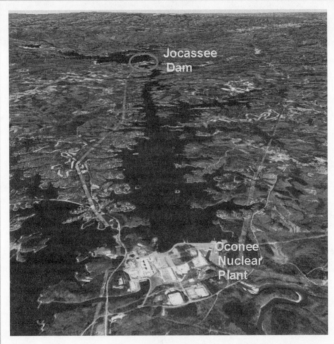

Thirty-four reactors at twenty sites around the United States are located downstream from large dams. The threat posed by the failure of those dams was not taken into account when the plants were licensed. One plant especially at risk is the three-unit Oconee Nuclear Station in South Carolina, which sits downstream of the nearby Jocassee Dam. More than 1.4 million people live within fifty miles of Oconee. *Google/Union of Concerned Scientists*

For years, the NRC and Oconee's owner, Duke Energy, have been at odds about the magnitude of risk to the plant from a failure of the nearby Jocassee Dam. Although the agency and the company disagreed about the risks there, they did agree on one thing: the consequences—a prolonged station blackout leading to core melting in less than ten hours and containment failure in less than three days. A "significant [radioactivity] dose to the public would result," according to Duke's own estimates.

Although the NRC had been aware of the problem since at least 1996, the agency had not required safety enhancements. NRC staff members themselves were divided on how to proceed, and the issue was still unresolved when Fukushima Daiichi demonstrated that prolonged station blackouts associated with flooding were more than a theoretical possibility.

Five days after the tsunami struck the Japanese plant, an NRC staff member e-mailed a colleague to inquire about the status of a Justification for Continued Operation (JCO) for the Oconee site, which was under dispute within the NRC. "In light of the recent developments in Japan," he wrote, "is anyone having second thoughts about the JCO, Oconee's path forward, the entire issue, etc.? Although the scope of such a disaster might be more limited at Oconee than in Japan—that is, the Japanese have other problems on their hands than a nuclear crisis, which is slowing them down to a degree, the Oconee disaster would be no less severe on the [reactor] units. Everything on site would be destroyed or useless."

In reply, his colleague wrote that the matter was being handed off to regional NRC officials to track as an "inspection project." "The tsunami should give management pause as the results of it sure look like what I would expect to happen to Oconee." Except, he noted, Oconee would receive more water.

Plans by Duke Energy to heighten a protective floodwall at Oconee, originally scheduled for completion in 2013, are now reportedly delayed until 2017.

Indeed, in spite of their proclamations to Senator Barrasso, the four commissioners had acknowledged that problems existed a few days earlier when they all voted to approve three major regulatory changes to address vulnerabilities identified by the NRC's own Fukushima NTTF. However, no longtime observer of the NRC expected the commissioners at the Senate hearing to openly profess doubts about the adequacy of U.S. reactor safety, even with memories of Fukushima Daiichi still vivid. The "it can't happen here" mindset is deeply rooted at the NRC, just as it was among Japan's nuclear establishment at the time of Fukushima. In fact, some would argue, the same mind-set has characterized the NRC's regulatory philosophy throughout its history.

The legacy of this mind-set cannot be undone without structural reforms.

But history holds out little hope that more fundamental changes can occur without a paradigm shift at the NRC. The record is replete with examples of the NRC staying the course even in the face of obvious warning signs.

A tortuous logic flourishes within the commission and influences its decisions. The NRC has always been reluctant to take actions that could call into question its previous judgments that nuclear plants were adequately safe. If it were to require new plants to meet higher safety standards than old ones, for example, the public might no longer accept having an "old one" next door. So the NRC is constantly engaged in an elusive quest for a middle ground from which it can direct needed improvements without having to concede that the plants were not already safe enough, thereby alarming the citizenry. It is not clear, even with the ruins of Fukushima in full view, that the NRC is willing to break out of that pattern.

In that regulatory balancing act, what has evolved over the years is a debate over "how safe is safe enough." In making its determinations on that issue, the NRC has all too often made choices that aligned with what the industry wanted but left gaping holes in the safety net.

There may be no better example than the long-standing controversy over the Mark I boiling water reactor design. Back in 1989, the NRC staff warned the commissioners that "Mark I containment integrity could be challenged by a large scale core melt accident, principally due to its smaller size." Staff experts recommended that the NRC require Mark I reactor owners to implement measures to reduce the risk of core damage and containment failure.

If the commissioners had taken effective action—action that would have sent a strong message to Mark I operators around the world, including those in Japan—it is quite possible that the worst consequences of Fukushima might have been avoided. Instead, the matter fell into a regulatory morass of competing interests and emerged with a resolution that accomplished little. It wasn't the first time that had happened.

In the aftermath of the 1979 Three Mile Island accident, the Kemeny Commission made clear that the safety status quo was inadequate. However, the panel explicitly refused to give its own views on "how safe is safe enough." Without any useful external guidance, the NRC embarked on a multi-decade struggle to provide an acceptable answer to this issue, the bane of regulators everywhere. It is without doubt a difficult public policy question, but the NRC's methods of addressing it have only created more confusion over the decades. Fukushima makes one thing clear: the process has not yielded the right answer.

In the NRC's world, the issue of "how safe is safe enough" is addressed through the concept of "adequate protection." When Congress created the agency out of the ashes of the Atomic Energy Commission, the NRC inherited the mandate bestowed upon its predecessor by the 1954 Atomic Energy Act: to "provide adequate protection of public health and safety." The NRC watered down this hazy concept even further by adopting a standard of "reasonable assurance of adequate protection" in its own guidance. The standard was so vague that it essentially gave the NRC and its five political-appointee commissioners a blank check for deciding exactly what constituted "adequate protection." In fact, several months after Three Mile Island, NRC chairman Joseph Hendrie said in a speech that "adequate protection means what the Commission says it means." *L'état, c'est moi.*

Even with such wide latitude, the NRC has always avoided specifying precisely what "reasonable assurance of adequate protection" means. This vagueness allows the commission to avoid drawing a line in the sand. In a 2011 speech, Commissioner Ostendorff, trained as a lawyer, emphasized that "reasonable assurance does not require objective criteria, and it is also determined on a case-by-case basis dependent on the specific circumstances." Although this policy preserved a great deal of flexibility for the NRC commissioners, the lack of a concrete standard injected a measure of subjectivity into NRC decisions that rendered them vulnerable to political shifts over the decades.

After the Three Mile Island fiasco, a key issue the NRC confronted was whether it could plausibly continue to maintain that its regulations provided "reasonable assurance of adequate protection." Certainly a lot of Americans didn't think so. But to admit that it had failed to meet that standard would have thrown into question the foundations of NRC regulations.

Up to that point, regulations were focused on a nuclear plant's ability to cope with the series of highly stylized events known as "design-basis accidents." On the NRC's list of design-basis accidents, the worst was one that assumed "a substantial meltdown of the core with subsequent release of appreciable quantities of fission products." But as catastrophic as that seems, it was far from the worst case. Owners did not have to consider the failure of more than one safety system at a time. They could assume that emergency core cooling systems would work, that pressure and temperature increases would be limited, and that core damage would be halted before the fuel could melt through the reactor vessel. And the containment structures needed only to be strong enough to prevent leakage of radioactive material to the environment under these modest accident conditions. (This requirement was typically met through the use of steel shells or liners. Reinforced concrete was used in containment buildings to keep things like tornado-driven objects from getting *into* the reactor, not to keep material within the reactor from getting *out*.)

Hypothetical events in which multiple safety systems failed, resulting in a complete core meltdown, failure of the reactor vessel, and breach or bypass of the containment, were considered by the NRC to be "beyond-design-basis" accidents. Prior to Three Mile Island, the agency deemed such events so improbable that, in contrast to its policy for design-basis accidents, "mitigation of their consequences [was] not necessary for public safety." At that time, to the NRC, to achieve "adequate protection" meant to protect against design-basis accidents.

To critics of the design-basis approach, Three Mile Island demonstrated its failure; to others, however, Three Mile Island represented a validation. The accident did not follow the design-basis script because multiple system failures occurred (owing to faulty equipment and human error); the core was severely damaged; and hydrogen exploded in the containment. However, those who saw the glass as half full pointed out that the accident was terminated before the core breached the reactor vessel; the containment never ruptured; and the amount of radioactive material that escaped was well below what the NRC considered acceptable for design-basis accidents.

In Three Mile Island's aftermath, one NRC-launched review of the accident came to a damning conclusion about this regulatory philosophy: "We have come far beyond the point at which the design-basis accident review approach is sufficient." In response, the NRC in October 1980 dutifully took up the question of whether it needed to amend its regulations "to determine to what extent commercial nuclear power plants should be designed to cope with reactor accidents beyond those considered in the current 'design basis accident' approach." To set that process in motion, it issued an Advance Notice of Proposed Rulemaking.[1]

In the Advance Notice, the NRC requested comment on numerous proposals for addressing the risk of beyond-design-basis accidents. These included requirements that containment structures be equipped with systems that could prevent them from being breached, such as filtered vents, hydrogen control measures, and *core catchers*, structures that could safely trap molten cores if they did manage to breach reactor vessels. Another suggestion was the addition of "an alternate . . . self-contained decay heat removal system to prevent degradation of the core or to cool a degraded core"—in other words, an external emergency backup cooling system with independent power and water supplies. And the NRC raised the possibility that reactor siting and emergency planning requirements might need to be tightened to address the greater radiation releases from beyond-design-basis accidents.

The Advance Notice of Proposed Rulemaking sent shudders through the nuclear industry. Companies feared that the NRC was setting the stage for a sweeping new rule that would require all plants to be able to withstand accidents previously considered beyond design basis, compelling the installation of costly new systems to cope with them. Without a clear boundary for "how safe is safe enough," such a regulation could open a Pandora's box of new requirements. There would be no telling how far it could go.

Members of the industry quickly united under the leadership of their U.S. trade association, the Atomic Industrial Forum (a predecessor to today's Nuclear Energy Institute), to head off the NRC by organizing a counter-campaign: the Industry Degraded Core Rulemaking program, or IDCOR. Funded with $15 million in contributions (more than $40 million in 2013 dollars) from nuclear utilities and vendors in the United States, Japan, Finland, and Sweden, IDCOR had as its goal to "assure that a rule, if developed, would be based on technical merits and would be acceptable to the nuclear industry."

IDCOR's extensive technical program included funding the development of a new computer code to simulate core melt accidents and support what were called "realistic, rather than conservative, engineering approaches." Yet there was little doubt what the program hoped to accomplish: to block any new regulatory requirements. While the NRC vacillated for four years on what new rules, if any, were needed, IDCOR marched toward its foregone conclusion. In late 1984, the group released its findings. The industry had drawn its own line in the sand. Risks to the public from severe accidents had been vastly overestimated; the actual risks were already so low that more regulation was not needed.

From IDCOR's perspective, even severe nuclear accidents posed little danger. That was because containment failure would take so long to occur that most fission products would have time to "plate out," or stick to structures within the damaged reactor, and would not be released to the environment. Therefore, the quantity and type of radioactive material that could escape during a severe accident—the source term—would be far below what the NRC had been assuming in its analysis of health impacts. In reality, no one would die from acute radiation exposure after even the most serious accident, and the numbers of cancer deaths would be hundreds of times smaller than previous studies had shown.

The industry's proposed reduction of the severe accident source term amounted to a bold *jujitsu* move to turn the NRC's original effort to strengthen regulations on its head. One requirement the industry was particularly anxious to undermine was the recently imposed ten-mile emergency evacuation zone around every nuclear plant. At the time, the evacuation requirements were causing a firestorm in New York State, where state and local authorities were blocking operation of the newly constructed Shoreham plant on Long Island by refusing to certify the evacuation plan. (Critics claimed the roads of narrow Long Island couldn't handle a mass exodus.) But if the amount of radiation that could escape the plant was so much smaller than previously believed, then perhaps a ten-mile evacuation zone wasn't needed.

The NRC made no attempt to hide its skepticism about the industry's source term recalibrations. At a 1983 conference, Robert Bernero, director of the agency's Office of Accident Source Term Programs, called those involved "snake oil salesmen." To help resolve the growing controversy, the NRC commissioned the American Physical Society, a respected professional association of physicists, to conduct a review of source term research. The physicists concluded that, although the evidence appeared to support reducing the assumed releases of certain radionuclides in certain accidents, there was no basis for the "sweeping generalization" made by IDCOR.

Ultimately, however, the industry's counter-campaign had an effect. Although the NRC refused to accept the industry's arguments, in 1985 the commission abandoned efforts to require protection against severe accidents and withdrew the Advance Notice of Proposed Rulemaking. In fact, the NRC went a step farther, issuing a Severe Accident Policy Statement that declared by fiat that "existing plants pose no undue risk to public health and safety." In other words, there was no need to raise the safety bar to include beyond-design-basis accidents because the NRC's rules already provided "reasonable assurance of adequate protection," the vague but legally sanctioned seal of approval. The NRC had already addressed Three Mile Island issues, and that was enough.

However, in the face of a growing body of research that suggested the safety picture was not quite that rosy, this declaration raised questions more than it provided answers. In the time-honored tradition of government bureaucracies, the NRC resolved to continue studying the issue, kicking the can farther down the road and confusing matters even more. While asserting that there were no generic beyond-design-basis issues at U.S. reactors, the commission held out the possibility that problems might exist at individual plants and that it should take steps to identify them. Even this proved controversial, requiring three years of give-and-take between the NRC and the industry merely to set ground rules for the study.

When the smoke cleared in 1988, the scope of the proposed Individual Plant Examination (IPE) program had been diminished to a mere request that plant owners inspect their own facilities for vulnerabilities to core melting or containment failure in an accident. What happened if the inspections actually turned up something was less clear. Even if the plant owners found problems, the NRC could not automatically require them to be fixed. The agency would have authority to do so only if such fixes represented "substantial safety enhancements" and were "cost-effective"—that is, if they passed the strict tests required by the NRC's recently revised backfit rule, which governed the changes it could require for existing plants.[2]

The 1988 backfit rule had its origins in the antiregulatory fervor of the Reagan administration. In 1981, President Ronald Reagan issued an executive order barring federal agencies from taking regulatory action "unless the potential benefits to society . . . outweigh the potential costs to society." Although such a cost-benefit analysis approach sounded reasonable to those seeking a way to reduce government interference, it was controversial for its coldly reductionist attempt to convert the value of human lives into dollar figures that could be directly compared to the costs incurred by regulated industries.

Although the NRC, like other independent agencies, was exempt from this executive order, a majority of commissioners wanted to adopt cost-benefit requirements anyway to add what they characterized as "discipline" to the backfitting process (as if the NRC were staffed by some sort of renegade regulatory militia.)

In the past, when the NRC had imposed new regulations, the industry complained that the resultant backfits were costly and often of little or no actual safety benefit. Proponents of cost-benefit analysis argued it would "address risks that are real and significant rather than hypothetical or remote." The key to this would lie in the use of sophisticated mathematical modeling to quantify risk. At the time of Reagan's executive order, the NRC's regulations only allowed it to impose backfits if they would "provide substantial, additional protection which is required for the public health and safety or the common defense and security." However, this standard was so vague that critics from both sides attacked it. Cost-benefit analysis in principle could help to solve that problem by providing a concrete, quantitative method for determining whether the benefits of a backfit—namely, the reduction in potential deaths or injuries following an accident—justified the costs.

Risk analysis had a receptive audience at the NRC. For many years, the NRC and its predecessor, the AEC, had engaged in a similar process. In the early 1970s, the AEC commissioned a pioneering project, the Reactor Safety Study, that attempted to use the tools of probabilistic risk assessment (PRA) to calculate the risk to members of the public of dying from acute radiation exposure or cancer as the result of a nuclear reactor accident. *Risk* was defined as the product of the likelihood of an occurrence and its consequences. One key conclusion was that even for nuclear accidents with very serious consequences, the "risk" each year to members of the public would be very low, since the probability of such accidents would be very low. That is, multiplying a large number by a very small number would yield a small number. The report, issued in 1975, famously came under blistering attack for its methodological problems and misleading implication that an average American had as much chance of being killed in a nuclear power plant accident as of being struck by a meteor.

One of the main criticisms of the Reactor Safety Study was that its calculations of probabilities "were so uncertain as to be virtually meaningless," as recounted by Princeton professor Frank von Hippel in his book *Citizen Scientist*. Each calculation required the input of thousands of variables, many of which had very large margins of error. If these uncertainties were not properly accounted for, the final result would be misleading. Consequently, many critics, including an independent review panel commissioned by the NRC, argued that probabilistic risk assessments were not precise enough to be used for calculating the absolute value of anything, particularly the probability that a given reactor might experience core damage in a given year.

A major source of PRA uncertainty is what types of events should be included in the calculation in the first place. Like good engineers, the early PRA practitioners began by analyzing things that they knew how to do—relatively well-defined events such as a pipe break. These are called *internal events* because they begin with problems occurring within the plant. But addressing *external events* like earthquakes, flooding, tornadoes, or even aircraft crashes proved more challenging. First of all, such events are notoriously hard to predict. Second, their consequences could be complex and difficult to model. Trying to come up with numerical values that would accurately describe the risks from these events was an exercise in futility. But instead of acknowledging that the failure to address external events introduced huge uncertainties in the nuclear accident risks they calculated, PRA analysts sometimes pretended that the possibilities didn't even exist—the scientific equivalent of reaching a verdict with crucial pieces of evidence missing.

Despite these technical challenges, the NRC eventually began to use PRA results more and more in its regulatory decisions—including the absolute values of accident risks that had been called "virtually meaningless." Over time, the agency began to view PRA risk numbers as more precise than they actually were. They were put to heavy use in the cost-benefit analyses that some commissioners wanted the NRC to rely on. That had a troubling consequence: as the risk of severe accidents appeared to shrink, so did the NRC's leverage to require plant improvements.

As if calculating the PRA risk values weren't complicated enough, cost-benefit analyses required another parameter to be specified: the monetary value of a human life. The NRC had carried such a number on its books since the mid-1970s: $1,000 per *person-rem*, a term used to characterize the total radiation dose to an affected group of people. Based on today's understanding of cancer risk, that put the value of a human life between $1 and $2 million. (The NRC failed to adjust for inflation for years, finally doubling the figure in the 1990s to about $3 million per life. That was about a half to a third the value placed on a human life by other federal agencies.)

Although a majority of commissioners embraced the cost-benefit approach, a key obstacle remained: could the NRC legally consider costs in making safety decisions? In the mid-1980s, the Union of Concerned Scientists and other public interest groups argued that the Atomic Energy Act did not allow cost to be considered at all; the NRC should base its decisions strictly on protecting public health. If a utility could not afford to build or operate a plant to meet that standard, then it would be out of luck. In response, the industry argued that the NRC had the right to consider the cost of backfits.

At first, the industry prevailed. In 1985, the NRC revised its rules to prohibit the commission from requiring any backfit unless it resulted in "a substantial increase in the overall protection of public health and safety . . . and that the direct and indirect costs of implementation . . . are justified in view of this increased protection." These tests were not required for backfits needed to fix an "undue risk," but the NRC refused to define what that meant.

Rather than simplify matters, the backfit test made them maddeningly unclear. In the end, it appeared that cost-benefit analyses would be required for essentially all proposed backfits, including any proposals for new regulatory requirements. Commissioner James K. Asselstine, a lawyer who headed the Senate investigation into the Three Mile Island accident before being appointed to the NRC, wrote a withering dissent to the new rule. "In adopting this backfitting rule, the Commission continues its inexorable march down the path toward non-regulation of the nuclear industry. . . . I can think of no other instance in which a regulatory agency has been so eager to stymie its own ability to carry out its responsibilities."

Asselstine, voting against the rule, contended that it imposed unreasonably high barriers to increasing safety and required a determination of risk "based on unreliable . . . analyses." He wasn't done: "The Commission also fails to deal with the huge uncertainties associated with the risk of nuclear reactors. The actual risks could be up to 100 times the value frequently picked by the Commission. . . . There is no reference in this rule . . . to how uncertainties are to be factored into safety decisions."

In 1987, the Union of Concerned Scientists, represented by attorneys Ellyn Weiss and Diane Curran, sued the NRC to block the rule, arguing that the commission could not legally consider costs in making backfit decisions. Later that year, an appeals court threw out the backfit rule, calling it "an exemplar of ambiguity and vagueness; indeed, we suspect that the Commission designed the rule to achieve this very result."

But the court's ruling created a peculiar two-tier system. In deciding *Union of Concerned Scientists v. U.S. Nuclear Regulatory Commission*, the court agreed that the Atomic Energy Act prohibited the NRC from considering costs in "setting the level of adequate protection" and required the NRC "to impose backfits, regardless of cost, on any plant that fails to meet this level." However, the ruling further confused the "how safe is safe enough" issue by concluding that "adequate protection . . . is not absolute protection." The NRC could consider the costs of backfits that would go beyond "adequate protection," the judges ruled.

The NRC revised the backfit rule accordingly in 1988. The court, by tying its decision to the largely arbitrary "adequate protection" standard, had preserved the agency's free hand to push safety in any direction it wanted. The NRC rebuffed calls to provide a definition of "adequate protection." The Union of Concerned Scientists failed to get the revised rule thrown out on appeal. Adequate protection would remain "what the Commission says it is."

The court's ruling essentially froze nuclear safety requirements at 1988 levels. If new information revealed safety vulnerabilities at operating plants, the NRC would have three options: conclude changes were needed to "ensure" adequate protection; redefine the meaning of "adequate protection" itself; or subject the proposed rules to the backfit test. (The NRC also kept a fourth option, an "administrative exemption," in its back pocket.) In any of these cases, most new safety proposals would have to leap a high—perhaps impossibly high—hurdle.

The new backfit rule threw a monkey wrench into the NRC's process for addressing severe accident risks. Because the NRC Severe Accident Policy Statement for the most part equated adequate protection with meeting the design basis, most new safety measures to deal with beyond-design-basis accidents were not needed for adequate protection. This meant that—unless the NRC were to admit that operating plants did not provide adequate protection, or to expand the definition of adequate protection, a step that could have major legal ramifications—it couldn't issue new requirements without showing that they were "substantial" safety enhancements *and* that they met the cost-benefit test.

When concern started to grow about the strength of the containment of one type of reactor in particular, the Mark I boiling water reactor, the NRC found itself in a straitjacket. If the agency were to require safety fixes for the Mark I containment on the basis of "ensuring" or "redefining" adequate protection, this would be seen as an admission that the fleet of Mark Is was unsafe. Otherwise, the NRC would have to prove the benefits justified the costs. That option would leave the fate of any safety improvements at the mercy of the risk assessors.

In the 1980s, there were twenty-four GE Mark I boiling water reactors in the United States. Because of their relatively small and weak "pressure suppression" containment structures, these reactors had been controversial almost from the time that the first commercial version, Oyster Creek in New Jersey, went on line in 1969. After the hydrogen explosion at Three Mile Island, in 1981 the NRC required that the relatively vulnerable Mark I and II containments be "inerted" with nitrogen gas to prevent such explosions.

But that was not the Mark I's only problem. As NRC staff members began to contemplate events they had never thought possible before Three Mile Island, additional frightening scenarios began to emerge. For instance, if a prolonged station blackout were to occur, operators would lose the ability not only to cool the core but also to remove heat from the containment, which could eventually over-pressurize and leak through seals not designed to withstand such high pressures and temperatures. Even worse, if power were not restored, the core would melt through both the reactor vessel and the steel containment liner. Such events would inexorably result in breaches of each of the multiple layers meant to prevent radioactive materials from reaching the environment.

The NRC was already addressing station blackout issues under a 1988 regulation. But that only required plants to develop a strategy to cope with a blackout for no more than sixteen hours. A prolonged station blackout at a Mark I reactor—one longer than the 1988 rule contemplated—would defeat the NRC's "defense-in-depth" multiple-barrier strategy for protecting the public. An NRC task force convened in 1988 to study the liner melt-through issue concluded that this vulnerability was a "risk outlier" that warranted prompt attention.

That was easier said than done, because the industry's IDCOR program was already out front with its opposite argument. At the same time the NRC's analyses were raising alarms about liner melt-through, IDCOR was asserting that the risks of core damage and containment breach were very low. In addition, an industry group, the Nuclear Utility Management & Resources Council (now folded into the Nuclear Energy Institute), immediately submitted a report opposing the NRC task force's conclusions, asserting that generic hardware modifications to the Mark I were not cost-beneficial because "the total risk from severe core melt accidents is low." In place of mandatory fixes to the Mark I, the industry wanted the NRC to consider plants on a case-by-case basis as part of the ongoing Individual Plant Examination program, which would take many more years to complete.

For the next several years, the NRC staff and the industry continued to wield dueling technical analyses to get the upper hand with the commissioners. To complicate matters, the NRC staff itself was divided, with some members aligning with the industry. The trade journal *Inside N.R.C.*, covering a three-day meeting in 1988 in which quarrelling NRC staff and officials were sequestered in a Baltimore hotel, quoted sources as stating that "there is literally a war going on" and alluding to instances in which "disagreement over the Mark I issue led to threats involving job security and research funding." One source told *Inside N.R.C.* that "there are some senior staff members who are doing everything they can to make sure the game is played by industry's rules . . . if the industry can win this one, they can win everything."

At the core of the dispute was the question of how much risk the Mark I fleet posed (an issue that would again loom large in 2011). Although the NRC's analyses showed a high likelihood that a Mark I reactor's containment would fail if the core were damaged, even the agency's staff believed that the chance of core damage was low—perhaps lower than at a pressurized water reactor like Three Mile Island. So the overall risk to the public might not be any greater from the Mark I than from other reactor types.

If so, it would be hard to demonstrate that fixing the Mark I problems would reduce risks by a big enough factor to satisfy the requirements of the backfit rule. And the majority of commissioners would not be likely to perturb the cherished meaning of "adequate protection" to address the problem via that route. A few weeks after the contentious Baltimore meeting, Themis Speis, the director of the NRC's research office, wrote an internal memorandum in which he contended that improvements to reduce the likelihood of containment failure would probably be blocked by the backfit rule.

Notwithstanding a likely defeat, the NRC staff went before the commissioners in early 1989 with a proposal for five critical improvements for Mark I plants that reduce their risk of core damage and containment failure. The staff asked the commissioners to:

- Speed up implementation of the 1988 blackout rule.
- Require acquisition of backup water supplies and pumps that could operate in a station blackout.
- Require hardened torus vents that could be used during accidents to expel steam and other gases to reduce containment pressure and temperature. Crucially, operators should be able to open and close the vent valves remotely and in the absence of AC power.
- Require that the systems needed to automatically depressurize the reactor vessel in an accident—an essential step to be able to pump emergency coolant into an overheating core—be made more reliable, especially in the case of an extended station blackout when the battery power needed to operate the valves would not be available.
- Require more robust emergency procedures to ensure that operators could effectively utilize all this new hardware.

The staff argued that certain Mark I containment failure modes, such as liner melt-through, could not be stopped should a core meltdown occur. The only strategy was to prevent meltdowns in the first place—and for that the backup coolant supplies and hardened containment vents were crucial. The staff presented calculations supporting its claim that the improvements would pass the backfit test. The staff's analysis showed that the owners of the Mark I could reduce the likelihood of core damage by a factor of ten by installing hardened vents: a substantial safety increase. The staff's calculations also showed the cost of the improvements justified the benefits.

At the commission meeting, the NRC staff faced a hostile audience in Commissioner Thomas Roberts and his colleagues. They were not helped by the fact that the Advisory Committee on Reactor Safeguards, the independent panel that reviews NRC activities, also vigorously opposed the staff and supported the industry.

Despite the commissioners' skepticism at the briefing, they remained deadlocked for months on the Mark I improvement proposal. The tiebreaker was Commissioner James Curtiss. The NRC staff told Curtiss that if the commissioners did not vote for the containment improvement program, and instead folded it into the Individual Plant Examination program, Mark I fixes would be put off for another five years. Curtiss apparently believed resolving the issue could not wait that long.

In July 1989, the commission finally made its decision. Although the final vote was 3–2 in favor of taking action, the outcome was far short of what the staff had requested. Of the five recommendations, the commission accepted only two, and even then it pulled its punches. First, it authorized speeding up the timetable for Mark I plants to comply with the station blackout rule—but this was an easy call, since it did not involve new requirements. Second, the commission decided to take action on hardened containment vents. But, reluctant to directly confront the industry on such a sensitive issue, the NRC gave Mark I owners an offer they couldn't refuse: install hardened containment vents voluntarily or the NRC staff would conduct plant-specific backfit analysis to determine if the agency could legally require them to comply.

The commission's offer presented an easy choice for most Mark I owners. If they installed the vents as a voluntary initiative, they would not have to submit a license amendment to the NRC for approval, and the NRC would have almost no regulatory control over the vents.[3] Although the NRC set basic standards for the design, construction, maintenance, and testing of the vents, Mark I operators would be under no obligation to meet them. The NRC would also have no authority to issue violation notices if it found problems, unless the vents interfered with other safety systems. In contrast, if the agency could show that the hardened vents passed the backfit test, it could force reactor owners to install and maintain them on the NRC's terms.

Initially, owners of all but five reactors decided to voluntarily install hardened vents. For the holdouts, the NRC followed through on its threat and did backfit analyses, concluding that it could require all five to install the vents. Four of the plant owners gave up at that point and "voluntarily" complied before they were forced to. The fifth owner—the New York State Power Authority—went on the offensive, challenging the staff's analyses and cost-benefit calculations regarding its James A. FitzPatrick nuclear plant. This time, it was the NRC's turn to buckle. FitzPatrick, located on Lake Ontario near the town of Oswego, became the only Mark I BWR in the United States that did not harden its vents.

The NRC staff audited some of the hardened vent designs and inspected the hardware and operating procedures after the vents were installed. But the fact that the licensees had performed the work voluntarily severely restricted the NRC's ability to ensure that the vents would be usable when needed.

And there was good reason to believe that they wouldn't be usable. The vents were designed to operate only within the design basis of the plant and only before core damage occurred. That meant that in the event of more severe conditions—high radiation fields, heat, or pressure—the vents might not function. And the NRC staff did not even require the vents to function during a station blackout, relegating that issue to future consideration in the Individual Plant Examination program. Once again, it was one step sideways.

As to the remaining three staff recommendations for Mark I improvements—alternate ways to inject water, improved reliability of reactor vessel depressurization, and emergency procedures and training—the majority of the commissioners supported the industry position that they should be folded into the quasi-voluntary IPE program. Accordingly, later in 1989 the NRC sent a letter to Mark I licensees meekly stating that the NRC "expects" them to "seriously consider these improvements during their Individual Plant Examinations."

This lackluster request received a lackluster response. When the NRC finally reported on the results of the IPEs in 1996, seven years after the project began, it noted that in several cases the licensees indicated that the containment performance improvements were being "considered, but do not identify the recommendations as commitments." Most of the Mark I plant owners stated that they already had alternate water sources and merely credited them in the IPE; some did not even bother to credit them. With regard to emergency training, the licensees simply committed to voluntary industry guidelines. And with regard to enhancing the reactor vessel depressurization system, many licensees did not respond at all. The NRC claimed victory in those cases when licensees actually did something, but it was powerless to compel any of them to do more; much less could it conduct thorough reviews and inspections to verify that what they had done would lead to meaningful safety improvements.

One could hardly judge the outcomes of the Mark I containment improvement program and the IPEs to be successes. Yet they set a major precedent for dealing with severe accident issues through "voluntary industry initiatives" (sometimes also confusingly called "regulatory commitments"). Like the backfit requirements of the 1980s, this was in keeping with the regulatory trends of the times, in which industry "self-regulation" tools like voluntary codes of conduct were increasingly used to forestall new government mandates, despite concerns about foxes guarding henhouses. For its part, the nuclear industry now could tout its voluntary actions as examples of its commitment to safety beyond what the NRC required.

One of the key voluntary industry initiatives of the 1990s was the development of Severe Accident Management Guidelines, or SAMGs. These were emergency plans plant operators were to use during an accident in which core damage had already occurred or was imminent. (SAMGs were to be used if a plant's emergency operating procedures, which in contrast were regulated by the NRC, failed to prevent core damage.) In 1994, the industry, under the auspices of its newly constituted advocacy group, the NEI, developed a guideline document that all licensees promised to adopt. Once again, however, because the SAMGs were voluntary practices, the NRC was virtually powerless to ensure that they would be workable and that plant workers would be appropriately trained to use them.[4]

Mark I containments were not the only ones that concerned the NRC; the Mark II had similar issues. Also, another type of containment—the Westinghouse ice condenser, a PWR version of a pressure-suppression containment—was vulnerable to failure in severe accidents, especially in the event of a hydrogen explosion. (Although it required the Mark I and II to be inerted with nitrogen gas, the NRC had not done so for ice condensers, or another model of BWR called the Mark III.) Several years after Three Mile Island, the NRC had required owners of ice condensers and Mark III plants to install igniters—similar to spark plugs—that could burn off hydrogen accumulating in a containment before it reached an explosive concentration. However, those igniters required AC power to function, so they wouldn't be available in a station blackout, a potentially major weakness. Even the gold standard—large, dry PWR containments—might be vulnerable to accidents in which the reactor vessel failed at high pressure. But the NRC's failure to impose meaningful changes on the Mark I, perhaps the worst of the lot, did not bode well for the future of the containment improvement program.

Over time, the NRC staff appeared to lose its appetite for grappling with the industry over new requirements to reduce severe accident risk. Even worse, in response to growing political pressure, the NRC decided to sweep other stubborn issues under the rug. In fact, as Three Mile Island receded into the past and no other Western-designed reactor experienced an event to jolt the memory (Chernobyl didn't really count, as it was considered an exotic Soviet beast), the agency in the 1990s embraced a sentiment that its requirements were not too lenient but rather too strict.

According to this line of thinking, severe accident risks were already so low that certain regulations could be weakened without significantly affecting safety. The NRC dubbed this approach "risk-informed regulation," and counted on probabilistic risk assessment data to justify what it euphemistically referred to as "reducing unnecessary conservatism" but actually amounted to removing safety requirements. Risk-informed regulation was seen by critics (such as David Lochbaum of the Union of Concerned Scientists) as a "single-edged sword": it was only used to reduce regulatory requirements, never to strengthen them.[5]

Reservations about the validity of probabilistic risk assessments faded as more and more utilities began to use them in regulatory applications. And why not? They seemed to enable the utilities to get what they wanted: less regulation. But even though PRA methodology had advanced, it still suffered from many of the same problems, including huge uncertainty factors when addressing earthquakes, other external events, and reactor shutdowns (when the risk of an accident can be surprisingly high). Again, the tendency of the NRC and plant owners when confronted with these uncertainties was to downplay or ignore them. As a result, safety decisions based on PRA analysis did not accurately account for the risks of these additional hazards. The misuse of PRA analysis did, however, lend credence to the concerns James Asselstine raised in his 1985 vote on the Severe Accident Policy Statement, when he accused his colleagues of deliberately ignoring uncertainties to minimize risks: "the Commission chooses to rely on a faulty [risk] number which supports the outcome they prefer."

Take the issue of developing a reliable PRA for an earthquake, which very few plant owners have done. To perform the assessment properly one would need accurate estimates of the likelihoods of earthquakes at each magnitude; detailed models of the effect that a quake of each magnitude would have on plant structures; and a defensible analysis of how the earthquake damage would affect plant operation and the ability of operators to carry out manual actions. Assembling this information would be a daunting task, and the uncertainties at every step would be formidable. There is little wonder that the industry has had difficulty tackling seismic PRAs.[6]

Among the first regulations that the NRC set its sights on "risk informing" was a post–Three Mile Island requirement that all reactors install "recombiners" that could prevent the accumulation of hydrogen during a loss-of-coolant accident. In reconsidering the requirement for the Mark I and II, the NRC's analysis found that the recombiners would not be needed to prevent hydrogen explosions during the first twenty-four hours after an accident because the reactor containments were inerted with nitrogen. However, the recombiners could be useful after twenty-four hours had passed because the inerting would become ineffective.[7] Nonetheless, in 2003 the NRC eliminated the recombiner requirement, concluding that removing this equipment would not be "risk significant." The reason: the SAMGs at those plants called for operators to vent or purge hydrogen in a severe accident, and the NRC believed that twenty-four hours gave them plenty of time to prepare to get that done. Based on this calculation, the agency concluded that the monetary value of the increased threat to public health was less than what the utilities would save by not having to maintain the recombiners—$36,000 per year per reactor.

Thus the NRC removed regulatory requirements to prevent hydrogen explosions in part by taking credit for voluntary initiatives—SAMGs—that it did not regulate. This type of twisted logic was typical of the risk analysis that enabled the NRC to weaken its regulations at the beginning of the twenty-first century.

In the years following Three Mile Island, the Japanese closely studied the NRC's regulatory reforms, and in many cases emulated them. Japan's Nuclear Safety Commission identified fifty-two lessons learned from Three Mile Island that it recommended for adoption in Japan's own safety regulations. Japan also began to develop severe accident countermeasures after Chernobyl. Among those that TEPCO incorporated at Fukushima were hardened vents, modifications to allow use of fire-protection pumps to cool the core if needed, and measures for coping with station blackouts of modest length, including loss of DC power.

The Japanese also developed severe accident guidelines, referred to as *accident management (AM) measures*, using the results of probabilistic risk assessments conducted by research organizations. In short, there were many similarities between actions taken in the United States and those in Japan.

Japan's severe accident management measures also shared many of the defects of the U.S. approach. All of the AM measures were rooted in the belief that the possibility of severe accidents was so low as not to be "realistic from an engineering viewpoint"; hence these steps were not considered essential. Consequently, the NSC concluded that "effective accident management should be developed by licensees on a voluntary basis," and the utilities accordingly developed AM measures on their own.

As a result, no regulator assessed whether the plant owners' assumptions were realistic regarding the ability of workers to carry out AM measures like hardened vent operation and alternate water injection. In particular, no one asked TEPCO why its AM procedures were designed to cope with a station blackout that would last only thirty minutes and affect only one reactor at a site. If someone had, perhaps TEPCO would not have had to concede after Fukushima that the tsunami and flood resulted in "a situation that was outside of the assumptions that were made to plan accident response."

Suppose that decades ago the NRC staff had succeeded in pushing through a much more aggressive approach for dealing with Mark I core damage and containment failure risks, including the challenges of a prolonged station blackout. There is no guarantee that the Japanese would have followed suit, but they would have been hard-pressed to ignore the NRC's example. The NRC staff in the 1980s had all but predicted that something like Fukushima was inevitable without the fixes it prescribed, but the agency's timidity—or perhaps even negligence—contributed to the global regulatory environment that made Fukushima possible. The NRC's reliance on the flawed assumption that severe accident risks are acceptably low helped to perpetuate a dangerous fallacy in the United States and abroad. Ultimately, the NRC must bear some responsibility for the tragedy that struck Japan. And the commissioners must acknowledge that unless they fully correct the flawed processes of the past, they cannot truthfully testify before Congress that a Fukushima-like event "can't happen here."

10

"THIS IS A CLOSED MEETING. RIGHT?"

It was the last session of the NRC's twenty-third Regulatory Information Conference. The RIC, as it is known, is an annual gathering that attracts regulators, utility executives, industry representatives, the media, and others for discussions of new and ongoing initiatives by the NRC. More than three thousand people from the United States and around the globe, including a team of seismic experts from Japan, had descended on a Marriott conference center across Rockville Pike from the NRC's White Flint headquarters for the three-day event.

Now, as the conference was winding down, a few dozen people had gathered to hear a panel discuss the latest results of an NRC research project entitled State-of-the-Art Reactor Consequence Analyses, or SOARCA, as it was called in the NRC's acronym-rich environment. The takeaway message from the panel: even if a severe nuclear power plant accident were to happen—say, an extended station blackout at a Mark I boiling water reactor—it wouldn't be all that bad.

The date was March 10, 2011.

By all accounts the RIC had been a great success, a reflection of how the NRC's stature had grown along with the improving fortunes of nuclear power in the United States. After decades without a new reactor order being placed, the United States in recent years had begun to see a resurgence of interest in nuclear energy, spurred on by policy makers, pundits, and industry boosters addressing a public that had largely forgotten the nuclear fears of three decades earlier. They argued that the atom was the only realistic alternative to greenhouse gas-belching fossil fuel plants for delivering large amounts of power to an increasingly energy-hungry world. That message was gaining traction, even among some longtime nuclear skeptics.

Nuclear energy's prospects were boosted by Congress in 2005. That year's Energy Policy Act (EPAct) contained energy production tax credits and loan guarantees to help insulate utility investors from the formidable financial risks that had crippled many past nuclear projects.

Thanks to incentives such as these, the NRC was soon besieged by more nuclear plant license applications than it could handle. To cope with the increase in its workload, the agency needed to expand significantly for the first time in decades.

By 2011, some of the momentum had been siphoned off by the persistent recession, which froze credit markets and reduced energy demand, as well as by the ultracheap natural gas made available by hydraulic fracturing. But the so-called nuclear renaissance was very much alive in the nation's capital. Interest remained high among many in Congress and within the Obama administration.[1]

Turnout for the 2011 RIC reflected the renewed support. The conference reported the highest attendance in its history, and sessions such as those devoted to the technology *du jour*—small modular reactors that could be installed in all sorts of unlikely places around the world—generated so much excitement that auditoriums filled to capacity and people were turned away at the doors.

As for the nagging issue of safety? That no longer seemed a showstopper, thanks in large measure to some deft messaging by the nuclear industry, led by the NEI. The long-ago accident at Three Mile Island represented the nuclear industry of old; the accident at Chernobyl was irrelevant to Western designed and operated nuclear plants. An entire generation of Americans had reached adulthood without encountering a major nuclear mishap. Perhaps things *had* changed when it came to nuclear safety.

This was all good news for the NRC.

Despite its official status as a neutral regulator, the NRC had been doing its part to promote the image of nuclear power as safe. SOARCA was a key element in that campaign. The goal was to supplant older NRC studies that estimated the radiological health consequences of severe reactor accidents. Many in the nuclear power community, both inside and outside the NRC, believed that those studies, dating back more than twenty years, grossly exaggerated the potential danger. Antinuclear groups were misusing old information to frighten the public, they argued. It was time for a new counteroffensive.

In the 1980s, the industry had asserted—via the findings of its own Industry Degraded Core Rulemaking (IDCOR) program—that the NRC was wildly overestimating the radiation releases that could result from nuclear accidents. At the time, the NRC staff, bolstered by the independent review of the American Physical Society, did not concur. However, times had changed. Now the NRC itself was leading the charge to reduce source terms. That gave rise to a new state-of-the-art study: SOARCA.

But in 2011, after spending five years and millions of dollars on the project, the NRC had a new problem: the numbers SOARCA was generating weren't cooperating with the safety message agenda. It was déjà vu for the NRC, which has grappled with how to explain away inconvenient facts about nuclear power risks over its history.

In the RIC's final hours, a panel of NRC experts clicked open their PowerPoint presentations and provided a SOARCA update. Only the most attentive would have noted a subtle change in the language describing the study's findings—an attempt, perhaps, to glide past some of SOARCA's unwelcome results.

No one in the room could know that these findings, the outcome of computer simulations, were about to be put to the test.

If the conference had taken place a month later, with Fukushima's devastated reactors still held in check by seawater while thousands of refugees lamented their poisoned homes, SOARCA's message would have been far different. The panelists would have known by then that the accident scenarios they had analyzed were no longer just theoretical constructs, but instead described real-world events with real-world consequences.

It was now clear that the release of even a small fraction of the radioactive material in a reactor core was enough to wreak havoc around the world and fundamentally disrupt the lives of tens of thousands of people. This was something SOARCA, designed to calculate only numbers of deaths, was not capable of predicting.

Nearly three decades earlier, in November 1982, Representative Edward Markey of Massachusetts held a press conference in Boston with Eric van Loon, executive director of the Union of Concerned Scientists, to disclose troubling information: the NRC was suppressing the results of a study that estimated the consequences for human health and the environment of severe accidents for every nuclear power plant site in the United States.

The NRC staff had drafted a report on the study for public consumption, but the commission had been sitting on it for over six months. At the time, three and a half years after Three Mile Island, antinuclear sentiments in the United States were running high. The possibility that the NRC was engaging in some sort of cover-up about risks confirmed the suspicions of many about the pronuclear bias of the agency.

The study, performed by Sandia National Laboratories and given the bland title "Technical Guidance for Siting Criteria Development," soon became known as the CRAC2 study, after the computer code it employed ("Calculation of Reactor Accident Consequences"). Among the calculations in the study was a projection of the dispersal of large radioactive plumes from a severe accident with containment failure and an estimation of the resulting casualties.

Like a civilian version of the models used by Cold War–era military strategists that ranked the outcomes of thermonuclear conflicts in impersonal terms like "megadeaths," CRAC2 was used to quantify the damage from nuclear accidents: the numbers of radiation injuries, "early" fatalities from high levels of radiation exposure, and "latent" cancer fatalities from lower and chronic exposures.

CRAC2 and other radiological assessment codes, like Japan's SPEEDI and the NRC's RASCAL, utilize complex models to estimate doses to individuals by simulating the way radioactive plumes released by a nuclear accident are transported through the atmosphere and the biosphere. CRAC2 went beyond those other codes by using more detailed models of the ways people could be exposed: external irradiation by radioisotopes in the air and on the ground, inhalation, and consumption of contaminated food and water.

CRAC2 also had a crude model for estimating the economic consequences associated with land contamination, addressing issues such as lost wages and relocation expenses for evacuees, and costs of cleanup or temporary condemnation of contaminated property. What it couldn't estimate were nonquantifiable consequences such as the psychological impacts on people forced to leave their contaminated homes and businesses either temporarily or permanently.

Radiological assessment codes like CRAC2 must incorporate many moving parts—plumes are traveling, radioactive particles are being deposited, and the population itself is not sitting still. Each calculation requires the input of hundreds of parameters, from source terms to types of building materials to the movement of evacuees and, eventually, even to the long-term effectiveness of decontamination efforts and the radiation protection standards governing people's return to their homes. Consequently, the results are very uncertain. Far too often, however, the tendency among regulators has been to endow these rough estimates with more authority than they deserve.

One of the largest sources of uncertainty is weather. Some types of weather could be much more hazardous than others, depending, for example, on whether the wind was blowing toward heavily populated areas and whether there was precipitation. But the NRC's as yet unreleased CRAC2 draft report contained only averages for weather conditions.

What Ed Markey and Eric van Loon presented to reporters that November day were not just the averages but the "worst case" results for the most unfavorable weather, such as a rainstorm washing out the plume as it passed over a large population center. In these projections, the "peak" early fatalities, as they were called, were far greater than the average values in the NRC report. The numbers were in fact shocking: for the Indian Point plant, thirty-five miles from midtown Manhattan, a worst-case accident could cause more than fifty thousand early fatalities from acute radiation syndrome. In contrast, the average value for early fatalities was 831.

The NRC had held on to the draft CRAC2 report for over half a year, presumably because officials worried that even the average-value casualty figures would be too much for the public to swallow. Markey's disclosure of the worst-case spreadsheet forced the NRC's hand, and it finally released the report that same day. The commission was quick to defend its decision not to include the worst-case results, offering a rationale that would become familiar over the years: the chances of an accident severe enough to produce such death and destruction were so slight as to be hardly worth mentioning. Or, as the NRC's head of risk analysis, Robert Bernero, said at the time, the likelihood of worst-case conditions was "less than the possibility of a jumbo jet crashing into a football stadium during the Superbowl."

For the next two decades, this line of reasoning formed the backbone of the NRC's strategy for addressing the threat of severe accidents—namely, that events threatening major harm to the public were so unlikely that they didn't need to be strictly regulated, a view shared by Japanese authorities and other members of the nuclear establishment worldwide.

In its risk assessments, the NRC was careful always to multiply high-consequence figures by tiny probabilities, ending up with small risk numbers. That way, instead of having to talk about thousands of cancer deaths from an accident, the NRC could provide reassuring-sounding risk values like one in one thousand per year. The NRC was so fixated on this point that it insisted that information about accident consequences also had to refer to probabilities.[2]

However, critics argued that the probability estimates were so uncertain— and there was so little real data to validate them—that the NRC could not actually prove that severe accidents were extremely unlikely. Therefore, accident consequences should be considered on their own terms.

In any event, the low-probability argument became less relevant in the aftermath of the September 11 aircraft attacks, when the public began to wonder what might have happened had al Qaeda decided to attack nuclear power plants that day instead of the World Trade Center and the Pentagon. No longer could one say with a straight face that a jumbo jet crashing into the Super Bowl was a one-in-a-billion event—if the pilot were intent on doing it deliberately. There was no credible way to calculate the probability of a terrorist attack and come up with a meaningful number. The NRC had long acknowledged this, and consequently did not incorporate terrorist attacks into its probabilistic risk assessments or cost-benefit analyses.

No longer able to hide behind its low-probability fig leaf, the NRC struggled to reassure Americans that they had nothing to fear from an attack on a nuclear power plant. While maintaining that nuclear reactors had multiple lines of defense, from robust containment buildings to highly trained operators, the NRC also had to concede that the reactors were not specifically designed to withstand direct hits from large commercial aircraft, and that it was not sure what would happen if such an attack occurred. The industry steered the public discussion toward the straw-man issue of whether or not the plane would penetrate the containment—it couldn't, according to the NEI—even though many experts pointed out that terrorists could cause a meltdown by targeting other sensitive parts of a plant.

To learn more about what could happen in an attack, the NRC commissioned a series of "vulnerability assessments" from the national laboratories, but the results remained largely classified for security reasons. Aside from a series of carefully constructed and vaguely reassuring talking points, the NRC provided few details beyond "Trust us." Communities near nuclear plants would get few tangible answers about the vulnerability of reactors in their midst.

Meanwhile, the 9/11 disaster had provided an opening for environmental and antinuclear groups to once again raise the safety concerns that had faded from view since Chernobyl. In the vacuum of new public information from the NRC, activists found ample fodder in the old CRAC2 study and its references to "peak fatality" and "peak injury" zones. Among them was the organization Hudson Riverkeeper, campaigning to shut down the Indian Point plant. Interpreting the CRAC2 results liberally, the organization's leader, Robert F. Kennedy Jr., spoke of dangers to the many millions of people within what he referred to as Indian Point's "kill zone."

Such talk was deeply upsetting to one NRC commissioner in particular: Edward McGaffigan. A voluble, intellectual, and pugnacious former diplomat and Senate defense aide, McGaffigan began his tenure on the NRC in 1996 by extending open channels of communication to the public. But after 9/11 he became openly hostile toward anyone he believed was exaggerating the dangers of nuclear power or misinterpreting the results of NRC technical studies.

"The media holds us to a very high standard, that what we say is factually true . . . but the antinuclear groups . . . basically get away with saying almost anything, however factually untrue it is," McGaffigan told a gathering of NRC staff in 2003, adding, "The way we fix it is we work aggressively to get our story out." The story, in his view, was that nuclear power was safe. Those who argued otherwise were misinformed and misguided.

McGaffigan was not alone in his frustration; other commissioners also accused critics of scare tactics. But McGaffigan went further, mocking members of the public who expressed the views he disdained.

McGaffigan's views worried nuclear watchdog groups. After all, how could a regulator be trusted to make the decisions necessary to protect public health if he had such absolute faith in the benign nature of the facilities he oversaw and did not worry about the effects of low-level radiation?

But there was more to McGaffigan's crusade; he accused the NRC staff itself of overstating the hazards of nuclear accidents. In his view, disinformation was coming from *inside* the agency as well as from hostile critics elsewhere. The staff's technical analyses, he said, were making unrealistically dire assumptions. One case in point was the risk posed by spent fuel pools, a subject that would surge to relevancy in little more than a decade at Fukushima Daiichi.

Edward McGaffigan, who was an NRC commissioner from 1996 to 2007. *U.S. Nuclear Regulatory Commission*

In January 2001, the NRC staff released a report, "Technical Study of Spent Fuel Pool Accident Risks at Decommissioning Nuclear Power Plants," or NUREG-1738. The report evaluated the potential consequences of an accident, such as a large earthquake, leading to the rapid draining of a spent fuel pool. NUREG-1738 estimated that within hours such an event could cause a zirconium fire throughout the pool that would result in melting of the fuel and release of a large fraction of its inventory of cesium-137 as well as significant amounts of other radionuclides. The study found that dozens of early fatalities and thousands of latent cancer fatalities could result among the downwind population.

Data from that study was later incorporated by outside experts into a technical paper, which was published in 2003 in the respected journal *Science and Global Security*.[3] The paper concluded that the U.S. practice of tightly packing spent fuel in pools was risky. It called on utilities to move most of the fuel to safer dry storage casks. In an angry response, McGaffigan called the publication, based at Princeton University, a "house journal" of "antinuclear activists." He fumed that "terrorists can't violate the laws of physics, but researchers can." But he also denounced NUREG-1738 itself, calling it "the worst" of excessively pessimistic staff studies on spent fuel vulnerabilities.

McGaffigan was so sure the Princeton study was wrong that, in a March 2003 public meeting, he appeared to direct the NRC staff to rebut the study before the staff had completed its own analysis. Such interference by a commissioner was practically unheard of. The NRC's inspector general investigated and concluded that McGaffigan had tried to exert inappropriate influence on the research staff.[4]

It was in this overheated political environment that the SOARCA study was conceived. The early results of the nuclear plant vulnerability assessments that the NRC had been conducting since shortly after the 9/11 attacks indicated, in the agency's view, that the radiological releases and public health consequences resulting from terrorist-caused meltdowns generally wouldn't be as catastrophic as previous studies, including CRAC2, had found.

Unfortunately for the NRC, it could not broadcast this good news because the vulnerability studies, being related to terrorist threats, were considered "classified" or "safeguards" information. But some inside the NRC reasoned that if the agency applied the same analysis methods to accidents instead of terrorist attacks, it might be able to dodge some of the security restrictions and get the information out to the public. SOARCA was the result.

There was a downside. Opening up the analytical process would also expose the staff's methodology and assumptions to unwelcome scrutiny by outsiders. So the NRC planned to keep a veil of secrecy over the SOARCA program itself, stamping the staff's proposal for how to conduct the study, as well as the commission's response, as "Official Use Only—Sensitive Internal Information." The NRC would control all information about the study and report the results only when it was ready, and in a manner that could not be—in its judgment—misinterpreted or misused. From the outset, one commissioner, Gregory Jaczko, objected, arguing that the study guidelines and other related documents should be publicly released. He was outvoted.

The NRC's concern about managing the information coming from SOARCA was evident from the beginning. The commissioners wanted the staff to develop "communication techniques" for presenting the "complex" results to the public. Although the technical analysis had barely begun, the first draft of the communications plan asserted that nuclear power plants were safe and had been getting safer for more than two decades. Even so, the commissioners rejected the draft and continued to micromanage the message. The communications plan would go through at least six revisions before they were satisfied.

One theme the NRC was determined to emphasize was SOARCA's scientific rigor. As the name suggested, the project was to be all about using "state-of-the-art information and computer modeling tools to develop best estimates of accident progression and . . . what radioactive material could potentially be released into the environment." But the NRC Office of Research, try as it might to be an independent scientific body, could never truly be free from the commission's policy objectives. The research office had faced accusations in the past of trying to influence the results of studies performed by its contractor personnel.[5] Now, the clear direction from McGaffigan and other senior officials would make it difficult to produce a completely objective study.

Although by all appearances the purpose of SOARCA was to reassure the public that nuclear power was safe, the nuclear industry did not enthusiastically jump on board. Perhaps company executives did not relish the prospect of another CRAC2-like spreadsheet making an appearance, listing potential accident casualty figures for every nuclear plant in the country—a recipe for bad publicity no matter how low the numbers. After all, Ed Markey was still in Congress, waging battles over nuclear safety.

The Nuclear Energy Institute interceded, sending the NRC a list of forty-four questions about the project, including a suggestion that a fictional plant be used instead of a real one. The SOARCA researchers soon found that very few utilities were interested in cooperating with the NRC on the study. (For added measure, the NEI hinted that any volunteers would want the right to review how their plants were portrayed.) The initial plan to analyze the entire U.S. nuclear fleet of sixty-seven plant sites was whittled down to eight and then to five; ultimately, only three were willing to participate. In the end, the NRC staff analyzed just two stations: Peach Bottom in Pennsylvania, a two-unit Mark I BWR, and Surry in Virginia, a two-unit PWR.[6]

With a vast, complex, and uncertainty-ridden study like SOARCA, it wasn't necessary to commit scientific fraud to guide the process to a desired outcome. There were plenty of dusty corners in the analysis where helpful assumptions could be made without drawing attention. The NRC employed a number of maneuvers to help ensure that the study would produce the results it wanted, selectively choosing criteria—in effect, scripting the accident.

It discarded accident sequences that were considered "too improbable," screening out events that would produce very large and rapid radiological releases, such as a large coolant pipe break. It only evaluated accidents involving a single reactor, even though some of the events it considered, such as earthquakes, could affect both units at either Peach Bottom or Surry. It considered its "best estimate" to be scenarios in which plant personnel would be able to "mitigate" severe accidents and prevent any radiological releases at all; it analyzed scenarios in which mitigation was unsuccessful but pronounced them unlikely. Perhaps most curious was the NRC's decision to assume that lower doses of radiation are not harmful—an assertion at odds not only with a broad scientific consensus but with the NRC's own regulatory guidelines.

The fog grew even thicker when the time came to decide how the study results would be presented. First, the commissioners decreed that figures such as the numbers of latent cancer fatalities caused by an accident should not appear. Instead, the report would provide only a figure diluted by dividing the total number of cancer deaths by the number of all people within a region. For instance, if the study predicted one hundred cancer deaths among a population of one million, the individual risk would be $100 \div 1,000,000$, or one in ten thousand. So rather than saying hundreds or even thousands of cancer deaths would result from an accident—guaranteed to grab a few headlines—the report would state a less alarming conclusion. And since the NRC's probabilistic risk assessment studies estimated that the chance of such an accident was only about one in one million per year, the current risk to an individual—probability times consequences—would be far less. To use the same example, it would be one million times smaller than one in ten thousand, or a mere one in ten billion per year: a number hardly worth contemplating. The communication strategy for SOARCA appeared to be taking its inspiration from the old Reactor Safety Study and its discredited comparisons of the risks of being killed by nuclear plant accidents versus meteor strikes.

But there was more obfuscation. The NRC would only reveal the values of these results for average weather conditions, and not the more extreme values for worst-case weather; this was the same strategy of evasion that had gotten the agency in hot water with Congressman Markey and the media back in the days of the CRAC2 report.

The commissioners also told the researchers to drop their original plans to include calculations of land contamination and the associated economic consequences. Earlier, the project staff had carried out such calculations for terrorist attacks at two reactor sites with high population density—Indian Point, north of New York City, and Limerick, northwest of Philadelphia—but apparently decided they did not want that kind of information to be made public. According to a staff memo, the models that had been used produced "excessively conservative" results—meaning, in NRC parlance, that the researchers thought the damage estimates were unrealistically high.[7] The staff said the models needed to be updated to obtain a "realistic calculation."

Some issues that emerged as the study progressed did not fit into the predetermined narrative. For instance, it was hard to explain why an earthquake or a major flood striking the Peach Bottom and Surry sites, each featuring side-by-side reactors, could be assumed to damage only one unit and leave the other unscathed. Logically, an accident involving both units would not only increase the source term, or amount of radioactive materials released to the environment, but also force operators to deal simultaneously with two damaged reactors. And, as the analysts noted, "a multiple-unit SBO [station blackout] may require more equipment, such as diesel-driven pumps and portable direct current generators, than what is currently available at most sites."

The analysts calculated that these scenarios had alarmingly high probabilities. But instead of following the study guidelines and including the scenarios, the staff decided in 2008 to recommend that the case of dual units be considered as a "generic issue"—a program where troublesome safety concerns are sent to languish unresolved for years. (The NRC was still pondering the recommendation three years later when Fukushima demonstrated that multiple-unit accidents were not merely a theoretical concern.)

The NRC's independent review group, the Advisory Committee on Reactor Safeguards, was not amused by what appeared to be a blatant attempt to bias the SOARCA study. In particular, it objected to the staff's seemingly arbitrary approach for choosing accident scenarios to analyze. It pointed out that SOARCA's good news safety message could be less the result of improved plant design or operation and more the result of "changes in the scope of the calculation." Simply put, SOARCA had analyzed different accident scenarios from those used in earlier studies like CRAC2, and therefore it could not be directly compared with them. Although the final SOARCA results might look better, that was because SOARCA was deliberately excluding the very events that could cause a large, fast-breaking radiation release of the kind CRAC2 had evaluated.

The NRC rebuffed its Advisory Committee's criticism and continued on the course the commissioners had set. In deference to public complaints that the study was being conducted in secret with no independent quality control, the NRC agreed to form a peer review committee. However, the NRC chose all the members, and the committee's meetings were also held in secret. The public would just have to trust that the committee was doing a good job.

When the NRC staff presented preliminary results of the study to the Advisory Committee in November 2007, it appeared that the staff had successfully obtained the conclusions its bosses wanted. First, the staff judged that all the identified scenarios could reasonably be mitigated—that is, plant workers, using B.5.b measures and severe accident management guidelines (SAMGs), would be able to stop core damage or block radiation releases from the plant. Even if they failed to prevent the accident from progressing, the news would not be too dreadful: the release of radioactive material would occur later and likely be much smaller than past studies had assumed, resulting in "significantly less severe" off-site health consequences. And finally, the NRC staff was so confident that it stopped the simulations after forty-eight hours, assuming that by then the situation would have been stabilized.

The results that the NRC staff presented to the Advisory Committee were striking. While CRAC2 had found that following a worst-case or "SST1" release,[8] acute radiation syndrome would kill ninety-two people at Peach Bottom and forty-five at Surry, SOARCA found the number of deaths to be exactly zero at both sites. There was no magic—or fundamental improvement in reactor safety—behind this stunning difference. The NRC had just fiddled with the clock. In the CRAC2 study, the radiation release began ninety minutes after the start of the accident, before most of the population within ten miles of the plant had time to evacuate, putting many more at risk. But the NRC had chosen accidents for SOARCA that unfolded more slowly. As a result, for most of the SOARCA scenarios, analysts assumed that the population within the ten-mile emergency planning zone would be long gone before any radiation was released. That way, by the time a release did occur, people would be too far away to receive a lethal dose, This was not an apples-to-apples comparison to the earlier study.

Harder to understand were the far lower numbers of cancer deaths projected by the SOARCA analysis, because even people beyond the ten-mile emergency planning zone could receive doses high enough to significantly increase their cancer risk. Whereas CRAC2 estimated 2,700 cancer deaths at Peach Bottom and 1,300 at Surry for this group, SOARCA project staff told the Advisory Committee that they had instead found twenty-five and zero cancer deaths, respectively.

That, too, involved sleight of hand—and some shopping around to find a convenient statistic. Despite a widespread scientific consensus that there is no safe level of radiation, the NRC staff decided to assume that such a level indeed existed: no cancers would develop until exposures reached five rem per year (or ten rem in a lifetime). Any exposure below that would be harmless.

At a 2007 Advisory Committee briefing closed to the public, Randy Sullivan, an emergency preparedness specialist for the NRC, let slip one reason why the SOARCA staff saw the need to use such an unconventional assumption. Apparently, the staff didn't like the numbers it would get if it used the widely endorsed linear no-threshold hypothesis (LNT), which assumes that any dose of radiation, no matter how low, has the potential to lead to cancer. It was an easy choice: assume a high threshold, predict many fewer cancers. Otherwise, the number of cancer deaths predicted by SOARCA would be so large it could frighten people.

At the briefing, Sullivan acknowledged: "We could easily do LNT, just go ahead, issue the source term, calculate it out to 1,000 miles, run it for four days, assess the consequences for, I don't know, 300 years and say 2 millirem times [the population within] 1,000 miles of Peach Bottom. What is that? Eighty million people. . . . We're going to kill whatever. This is a closed meeting. Right? I hope you don't mind the drama.

"So then we'll say that our best estimate is that there will be many, many thousands . . . you'll have 2 millirem times 80 million people and you'll claim that you're going to kill a bunch of them."

Considering a five rem per year threshold ultimately proved too misleading even for the SOARCA team itself to tolerate, so it eventually evaluated a range of thresholds, including the LNT assumption of zero. But other optimistic assumptions enabled the team to keep the numbers small. A 2009 update to the commissioners informed them that the study continued to find off-site health consequences "dramatically smaller" than those projected by CRAC2.

SOARCA was supposed to be a three-year project. But by the time of the SOARCA session at the March 2011 Regulatory Information Conference it had dragged on for nearly six years. (Commissioner McGaffigan would not live to see the fruits of the project's labors—he died in 2007.) Addressing the methodological problems that the Advisory Committee had criticized, running new analyses requested by the peer review panel, coping with problems with contractors, and project mismanagement all contributed to repeated postponements of the completion date. But perhaps the biggest time sink was the need to address more than one thousand comments from other NRC staff, who also had trouble swallowing some of the SOARCA methodology.

The unanticipated volume of staff comments was a clear indication of internal discomfort with the study. In January 2011, after being informed of yet another delay requested by the staff, Office of Research director Brian Sheron wrote in an e-mail, "[I]f we miss this date, I suggest we all start updating o[u]r resumes."

One of the major internal disagreements had to do with SOARCA's assumptions regarding so-called mitigated scenarios—in plain English, how fast and successfully operators could use the emergency tools at hand to wrestle an accident to a safe conclusion and avoid a radiation release. Could workers really start and operate the RCIC system at Peach Bottom without generator or battery power, as the SOARCA project staff had confidently concluded? Could they hook up and run portable pumps and generators to run safety systems for forty-eight hours (the limit of the SOARCA analysis)?

As early as 2007, project contractors at Sandia National Laboratories, which was reviewing the SOARCA analysis, were questioning whether all the emergency equipment and procedures would perform as the NRC team predicted. Sandia wanted what it called a "human reliability analysis."

That year Shawn Burns, a senior technical staff member at Sandia, wrote what proved to be a rather prescient letter to the NRC:

> The principal initiating event for the Long Term Site Blackout [at Peach Bottom] ... is a seismic event of sufficiently large magnitude ... to cause massive and distributed structural failures ... the realism of relocating relatively large and heavy mitigation equipment ... from their storage location(s) through rubble and other obstacles to their connection points in the plant is difficult to support. Similar questions would apply to the other plausible initiating events, including massive internal flood or large internal fire.

He went on to raise questions about other potential difficulties, including un-availability of backup cooling water supplies, electrical connection problems, difficulties with instrumentation, and dead batteries.

But the NRC commissioners had already spoken on this issue: emergency strategies and equipment—the so-called B.5.b. measures—would work in the SOARCA scenarios. Whether this was a reasonable assumption did not seem to figure into their instructions.

The issue bothered many within the ranks of the NRC, however. Computer models were one thing; *actual* hands-on experience at the nation's reactors was an entirely different matter. Senior reactor analysts who work in the NRC's regional offices for the Office of Nuclear Reactor Regulation, and who have all had previous experience as inspectors in the field, had a less optimistic view of the feasibility of these measures than the researchers running computer models at NRC headquarters. The reactor analysts had seen the B.5.b equipment for themselves.

The Advisory Committee on Reactor Safeguards also expressed doubts about the way the SOARCA report seemed to take the success of the B.5.b mitigation measures for granted. The committee asked whether the project staff had actually "walked down" the emergency measures to determine if they'd work under extreme accident conditions. The answer was no. Instead, the staff had based its conclusions on so-called tabletop demonstrations—that is, moving pieces around a toy model of each plant.

To quiet the skeptics, SOARCA staff visited Peach Bottom in Pennsylvania and the Surry plant in Virginia and examined the actual equipment. Internal dissent continued, but the SOARCA staff went on to question the validity of the concerns of critics and conclude that, based on those walkdowns at the plants, the likelihood of everything working was even greater than previously thought.

It was easy to understand why the project team wanted to believe that plant workers had the ability to mitigate the severe accidents that were ana-lyzed: everything else led to core meltdowns—albeit more slowly than previous studies had found.

An example was the SOARCA staff's analysis of a hypothetical "long term" station blackout at Peach Bottom, which is located about forty-five miles from Baltimore and eighty-five miles from Washington, DC. The accident scenario proceeded through a grim sequence of events. First, all electrically powered coolant pumps would stop working. Using batteries, operators could start up the steam-powered RCIC system, but after four hours, the batteries would fail, and after another hour, so would the RCIC. The temperature and pressure within the reactor vessel would quickly rise, and the safety relief valves on the vessel would eventually stick open, steadily releasing steam. With no makeup water available to replace the steam, the fuel would be uncovered in a matter of minutes.

At about nine hours, the fuel would start to melt, eventually collapsing and falling to the bottom of the reactor vessel. After about twenty hours, the molten fuel would breach the vessel bottom and spill onto the containment floor, where it would spread out and rapidly melt its way through the steel containment liner. A few minutes later, hydrogen leaking from the containment into the reactor building would cause an explosion, opening up the refueling bay blowout panels and blowing apart the building's roof.

If batteries were not available for those first four hours—if they were flooded from the adjacent Susquehanna, for instance—the resulting "short term" station blackout would be even worse. In that case, the models predicted that core damage would start after one hour and the containment would fail at eight hours.

In either case, once the containment failed a plume of radioactivity would escape the damaged plant and overspread the area. As expected, the calculations predicted no early fatalities from acute radiation syndrome. But the latent cancer fatality numbers told another story. Early SOARCA data supposedly had showed that health consequences were "dramatically smaller" than those predicted by CRAC2. After the SOARCA staff was repeatedly criticized for making assertions like these without ensuring that the two studies were consistently compared, analysts ran new calculations to see what would happen if SOARCA's methodology was applied to a CRAC2-sized release at Peach Bottom and compared to the SOARCA result for a station blackout. The outcome? The differences in latent cancer risk were not that dramatic.[9]

Within fifty miles of Peach Bottom, the estimated number of cancer deaths caused by a short-term station blackout, averaged over weather variations, would be 1,000, compared to the 2,500 projected by CRAC2 for a much larger radiation release. From a statistical standpoint, given the uncertainties, the difference was meaningless. And from a human standpoint, 1,000 deaths, while less than 2,500, was still a pretty unacceptable health consequence. Rather than discrediting the old CRAC2 analysis, the SOARCA study had in important respects validated it.[10]

For the small crowd gathered at the Marriott on March 10, 2011, to hear an update on SOARCA, some key details were missing. The NRC was still unwilling to release the numbers publicly. Anyone who wanted to know what was going on had to resort to Kremlinology to interpret the subtle changes in the statements that the NRC approved for release.

Two years earlier, at the last SOARCA session at the RIC, Jason Schaperow of the study team had reported a preliminary conclusion that "releases are dramatically smaller and delayed" from those projected in CRAC2, news that hardly surprised anyone in the room who had been watching the data contortions over the years. But in 2011, the statement had subtly changed. Patricia Santiago, the latest of several branch chiefs to oversee the SOARCA study, presented a bullet point that "for cases assumed to proceed unmitigated, accidents progress more slowly and *usually* [emphasis added] result in smaller and more delayed radiological releases than previously assumed/predicted."

The word *usually* was key here. It left the door open to the possibility that there were scenarios in which CRAC2's predictions had not been so far off base after all: perhaps, for instance, a station blackout at Peach Bottom.

Nor did Santiago repeat the now familiar statement about dramatically smaller numbers of cancer deaths. She noted only that "individual latent cancer risk for selected scenarios generally comes from population returning home after [the] event is over." In other words, most of the dose to the population would be received not during the early stages of the accident by people exposed to radioactive plumes, but long after the accident was over by evacuees who had no choice but to return to their now contaminated homes to live, hardly a consolation.

However, Santiago did highlight one SOARCA conclusion that had not changed over the years: the accident scenarios analyzed as part of the study "could reasonably be mitigated, either preventing core damage or delaying/reducing the radiation release."

The next morning, all the tabletop models and computer runs and fingers-crossed assumptions that supported that conclusion would face their first real-world test. At 11:40 a.m. on March 11, Jason Schaperow sent an e-mail to Santiago, his supervisor. "Today's Japanese earthquake seems to have caused one of the SOARCA scenarios (long-term station blackout)."

"On this morning's news they said no release," Santiago replied. "Time will tell."

Computer modelers and analysts love to obtain real-time data that they can use to validate the predictions of their models—except, perhaps, if they are simulating disasters. Soon, the SOARCA team would watch as many of the catastrophic events they had deemed improbable unfolded not on a computer screen, but on a television screen. And in the process the limitations of the SOARCA approach—the project in which the NRC had invested so much time and money to win over a skeptical public—would become evident.

Over the years as the SOARCA study progressed, it had revealed the potential for a natural disaster to cause a truly horrific event: an accident that involved multiple reactors, rendered most emergency equipment useless, and contaminated large areas with radiation plumes far beyond emergency planning zones due in part to the vagaries of weather. Yet instead of taking action to prevent such an accident, the NRC convinced itself that even if the accident did happen, the consequences would be minor. Difficult issues were disregarded or put off for another day.

If the NRC had undertaken this study not as an exercise in reinforcing existing biases, but as a roadmap for identifying and fixing safety weaknesses from America to Japan, SOARCA's most dire predictions might not have made the transition from a PowerPoint presentation to an event that shocked the world.

11

2012: "THE GOVERNMENT OWES THE PUBLIC A CLEAR AND CONVINCING ANSWER"

On January 18, 2012, protesters attempted to gain access to a closed-door meeting of Japan's Nuclear and Industrial Safety Agency (NISA). More than a hundred uniformed and plainclothes officers arrived to quell the scuffle. A green and yellow sign held aloft by one of the well-dressed demonstrators read: "Nuclear Power? Sayonara."

Before Fukushima Daiichi, that prospect seemed unlikely. Japan was firmly wedded to nuclear power—economically, politically, and socially. But now, as the consequences of the accident ten months earlier continued to unfold, to many Japanese the marriage was difficult to defend. The demonstrators being jostled in a crowded hallway and chanting "Shame on you" to the authorities sequestered behind the doors were part of a new, vocal constituency intent on having a say on Japan's energy future. For them, the price of nuclear power was just too high.

Since the accident, much had changed across Japan; 2012 would become a year marked by protests, uncertainty, and shifting political allegiances. Although Fukushima Daiichi officially was in cold shutdown, the popular debate over nuclear power was increasingly active. On numerous occasions, public opinion boiled over at politicians who seemed tone deaf to the concerns of average citizens. The country's leaders—in government and in the clannish private sector—appeared determined to close the book on the disaster of March 11 as quickly as possible and begin restoring the role of nuclear power as an essential element in Japan's economic life.

Immediately upon taking office in September 2011, Prime Minister Yoshihiko Noda had made his views clear: Japan needed nuclear energy. Noda proposed a middle ground—one that he hoped would ease public fears about living *with* nuclear power and corporate Japan's fears of living *without* it.

"[I]t is unproductive to grasp nuclear power as a dichotomy between 'zero nuclear power' and 'promotion,'" said Noda. "In the mid- to long-term, we must aim to move in the direction of reducing our dependence on nuclear power generation as much as possible. At the same time, however, we will restart operations at nuclear power stations following regular inspections, for which safety has been thoroughly verified and confirmed." If the public could be convinced that Japan's reactors were safe, perhaps the marriage could be saved.

The "how safe is safe enough" threshold for many Japanese had soared in the months since March 11, however. In normal times, safety inspections had been largely pro forma, conducted out of view, with little or no public input. Everybody had seen what that produced. If Noda thought his proposal to restart the nation's reactors would win public support, he quickly discovered otherwise.

Within days, tens of thousands of protesters took to the streets of Tokyo. The September 19, 2011, protest was the largest public demonstration in years in the capital. They wanted every reactor in Japan shut down permanently. A sign held by an elderly protester depicted a mother cradling an infant. It read: "This child doesn't need nuclear power."

On September 19, 2011, tens of thousands of protesters clogged the streets of Tokyo, demanding an end to nuclear power in the country. Here protestors gather at the Meiji Shrine Outer Garden in the nation's capital. *Wikipedia*

• • •

After being escorted out of their meeting room to avoid the noisy protestors, NISA officials reconvened and approved the restart of two reactors at
the Ohi nuclear plant in Fukui Prefecture, on Japan's west coast. At this point,
only three of the country's fifty-four reactors were still operating, and those
would also soon be turned off. Before any could be restarted, all were required to pass the safety checks first ordered by former prime minister Kan
and now cited by Noda: a first round of government-ordered "stress tests."
Plants that passed the first-round tests would then undergo a second, more
comprehensive round that would determine whether plants were safe enough
to continue to operate.

The first-stage tests were designed to verify a reactor's ability to withstand specific events, such as beyond-design-basis earthquakes, tsunamis,
or both. Taking a lead from the European Union, which ordered stress tests
for all one hundred thirty operating plants within its jurisdiction after the
March 11 accident, the Japanese government directed the nation's utilities to
perform similar tests. However, the guidelines were vague, failing to specify
clearly how far beyond the design basis the assumed stresses should go. That
made the outcomes of the tests hard to interpret.

The results of the first-stage tests were submitted to NISA, which then
passed on its plant restart recommendations to the Nuclear Safety Commission. The commission, in turn, forwarded its decision to the prime minister.
Noda and several of his cabinet members would have the final say.

Even as the government proceeded through the formalities of the stress
tests, pressure was growing from several influential constituencies to bring
the reactors back online as rapidly as possible. A recurrent theme among
the restart supporters was that Japan was headed for a major electricity shortage. That moment was approaching, along with the hot, humid months,
when energy use normally would rise by about 50 percent. Unless Ohi Units 3
and 4—the first reactors in line in the stress test review process—were restarted, Japan would be nuclear free on May 6, 2012.

That prospect posed a worrisome "what if?" for restart advocates. Suppose
Japan successfully made it through the summer without nuclear power. That
would make it more difficult to win public support for returning the plants
to service.

To add credibility to NISA's conclusion that the Ohi reactors were safe,
the Noda government invited a delegation from the IAEA to review the findings. Based on a preliminary assessment, the IAEA reported that the stress
tests were "generally consistent" with its own safety standards and supported

"enhanced confidence" in the reactors' ability to withstand even a disaster similar to that of March 11.

Those findings drew a scathing public rebuke from two nuclear experts who served as NISA advisors. The tests, the two men said, failed to take into account complex accident scenarios as well as critical factors such as human error, design flaws, or aging equipment. At a news conference, Masashi Goto, a former reactor designer, labeled the stress tests "nothing but an optimistic desk simulation based on the assumption that everything will happen exactly as assumed." [1] Goto also argued that the stress test methods should first be applied to Fukushima Daiichi to see if they could correctly simulate the outcome of the accident.

The other expert, Hiromitsu Ino, a professor at the University of Tokyo, accused the IAEA of simply rubberstamping NISA's decision because of the international agency's dual role as regulator and promoter of nuclear power. Ino noted that the IAEA had deemed the Kashiwazaki-Kariwa nuclear plant safe after a 2007 earthquake without conducting a full examination of the plant.

Ino's and Goto's sharp critiques were echoed two weeks later by another expert: Haruki Madarame, the head of Japan's NSC, who had been at Prime Minister Kan's side during the height of the accident. In a declaration that must have surprised many, Madarame asserted forcefully that Japan's nuclear regulatory framework was flawed, out of date, and below international standards. He claimed that the government's foremost concern was not protecting the public but promoting nuclear energy. Madarame added that Japan had become overly confident in its technical superiority, failing to acknowledge the risks that nuclear power posed, especially in an earthquake-prone country. Concerning the stress test results, he said: "I hope there is an evaluation of more realistic actual figures."

The IAEA's final review of Japan's stress tests, released in March, appeared to support some of the critics' concerns. The agency said that NISA had not communicated what its desired safety level was or how the assessments could demonstrate it. Regardless, the IAEA's overall judgment did not change.

For many Japanese, the epiphanies of such once-staunch nuclear proponents as Kan and Madarame only heightened uncertainty and confusion. Before March 11, 2011, they had been led to believe that their country's reliance on nuclear power was a prudent choice. Now, knowledgeable insiders were painting the entire nuclear framework as a public threat and an embarrassing fraud. And at the same time, the leadership in Tokyo was pushing to restart reactors.

If criticism from Japan's own experts wasn't unsettling enough, international disapprobation soon followed. A group of experts summoned by Yotaro Hatamura's investigative committee, which was wrapping up its lengthy assessment of the accident, weighed in. Because the Hatamura panel focused on TEPCO and events at Fukushima Daiichi, much of the criticism was directed at the utility, but the message spilled over: the failings were systemic. For example, Richard Meserve, former chairman of the U.S. NRC and president of the Carnegie Institution for Science, said that not only had TEPCO become overly confident but Japanese regulators had followed suit, falling victim to the myth of safety. "There has to be a willingness to acknowledge that accidents can happen," said Meserve.[2]

At the moment TEPCO was in a fight for its own survival. A company that many believed was too big to fail would soon become a ward of the state. This was more than just a corporate icon falling on hard times. Soon, the cost of TEPCO's lax oversight, failed planning, and propensity for high-stakes gambles would shift from a private debt to a massive public one.

Estimates of the cost of cleanup and compensation to victims of the Fukushima Daiichi accident were pegged at more than $71 billion (6.6 trillion yen) by late 2012. Much of that burden will fall on the shoulders of Japanese taxpayers, already suffering from a deeply troubled economy.

As Japan marked the first anniversary of the Fukushima Daiichi accident, the magnitude of TEPCO's financial woes was growing. The plunge in the company's stock price had erased almost $39 billion in market value, according to one analysis. Without an infusion of cash, the utility—which employed 38,000 people and supplied electricity to about 45 million, including those living and working in the world's largest city—almost certainly would collapse.

Private lenders and institutional investors were wary, wanting the government to craft a rescue. No one could say with any certainty just how much this all was going to cost. Rumors that the government was going to nationalize TEPCO were initially rebuffed by Trade Minister Yukio Edano. But an intervention of some sort increasingly seemed to be the government's option of choice—and TEPCO's only hope.

The utility's leaders remained defiant, asking Tokyo for money but rejecting any government say in company operations. Leading the fight to keep TEPCO under the control of insiders was its chairman, Tsunehisa Katsumata, one of Japan's best-connected business leaders. Although Katsumata planned to step down in a show of responsibility for the accident, he argued that only TEPCO executives knew how to run the company. Edano countered that

TEPCO's corporate culture "hasn't changed at all" and that new management was necessary.

Some economists and energy experts saw TEPCO's dire predicament as a golden opportunity to restructure Japan's antiquated energy supply system by breaking up utility monopolies and introducing competition. "Since the 1980s the utilities have looked more like bloated government departments than red-blooded businesses," a writer for *The Economist* noted.

Unlike energy markets in the United States and elsewhere, in which generation and distribution had been separated, opening the way for competition, in Japan the ten regional power companies held total monopolies on all aspects of energy supply. The result was reliable service but at a very high price. The monopolies were extremely lucrative for the utilities.[3] An analysis by Bloomberg News found that the ten of them earned $190 billion in annual revenues from generating, transmitting, and distributing power. Shattering that stranglehold, some Japanese academics and international analysts believed, could open markets, reduce rates, give renewable energy a toehold, and possibly lessen Japan's need for nuclear energy.

To critics of the status quo, the months following the accident had provided some hope that change was possible. During the summer of 2011, the government had imposed power restrictions on large corporate users. Voluntary conservation efforts had been surprisingly successful. Electricity consumption during steamy August had dropped by about 11 percent. The usual chill of air conditioning was gone, as were coats and ties. Some manufacturing operations moved to off-peak nights and weekends. Even some night baseball games were switched to daytime. That led to speculation that reduced consumption could become the norm and there would be no need to restart the reactors. The nation stood at a crossroads, activists argued, needing only visionary leadership to map out a new energy policy.

But those hopes gradually faded as it became clear that Japan, still struggling to recover not only from a nuclear disaster but from a massive natural disaster, was unable or unwilling to take on such a challenge. Restructuring the electricity distribution system was not a priority. Keeping TEPCO alive was, however.

In late March 2012, TEPCO requested $12 billion (1 trillion yen) from the government to stay afloat. The price, as company officials knew, was the end of their independence; the government would take over control of the utility.[4] Breaking with the utility's long tradition of promoting executives from inside, the government brought in Kazuhiko Shimokobe, a bankruptcy lawyer and turnaround specialist, to be TEPCO's chairman. Shimokobe, who had been running the government-backed TEPCO bailout fund, would handpick the new president. "This is the last chance to restore TEPCO," he grimly told reporters.

The ten-year bailout plan approved by the trade ministry in early May provided $12.5 billion in government capital in exchange for more than 50 percent of the voting shares of TEPCO, with the right to increase that percentage to two-thirds if the utility failed to institute major reforms. In addition, TEPCO had to agree to a cost-cutting plan and other changes. All sixteen directors would resign and be replaced by a new board, the majority of whose members would be from outside TEPCO. The new president would be not an outsider, as some had hoped, but Naomi Hirose, a TEPCO managing director who held an MBA from Yale University.

TEPCO's new management had to face a formidable challenge: a $9.78 billion loss for the fiscal year that ended in March 2012. A key step to move the utility back toward solvency was the restart of its seven reactors at the Kashiwazaki-Kariwa plant, three of which had been idle since the 2007 earthquake. "[A]ny plan that does not include nuclear energy would be nothing more than a pie-in-the-sky blueprint," said Shimokobe, who along with Hirose was slated to officially take over following shareholder approval in late June.

By late June, when Japan's power companies held their annual meetings, it was evident that TEPCO wasn't the only utility opposed to a nuclear phase-out. Shareholder proposals to reject nuclear power were soundly voted down at all the meetings. Unsurprisingly, the utilities' main institutional investors—life insurers and banks—voted with the company managements on this issue. The status quo seemed to them a safer bet. In the end, stockholders approved the bailout plan and the ongoing commitment to nuclear power.

Indeed, the recovery plan barely mentioned any energy strategies beyond restarting the utility's nuclear plants. The influential *Asahi Shimbun* editorialized that TEPCO was getting off too easily. The utility should be forced into bankruptcy and the company's shareholders and creditors should be "made to pay the price for their own mistakes." Only then would it be right to tap the Japanese public for help in a recovery.

As of May 6, 2012, Japan had no operating reactors. Without them, the system was shy of roughly fifty gigawatts of generating capacity—about what Tokyo consumes during peak periods. With no nuclear power in the mix, Japan would require an additional three hundred thousand barrels of oil every day, along with an annual import of 23 billion cubic meters of natural gas to meet electricity demand, according to International Energy Agency estimates. (The cost of this imported fuel was expected to more than double residential electricity bills across Japan and further hobble the nation's troubled economy.)[5]

Although the media had spent months speculating about a Japan without nuclear power, others never doubted that the reactors would eventually restart, given the depth of institutional support for nuclear energy. On March 23, the Nuclear Safety Commission ruled that the stress tests performed on Ohi Units 3 and 4 were satisfactory. And on April 13, Noda announced that the two Ohi reactors could safely resume operation. All that remained was the approval of local officials, no longer a sure thing for those who saw the devastation Fukushima had brought to nearby communities.

This rush to fire up the reactors made no sense to many in Japan. The accident at Fukushima Daiichi was still being investigated and many critical questions remained unanswered. The current regulatory system had clearly proven itself inadequate, yet the creation of a new agency to replace the discredited NISA was tied up by political infighting. For now, safety matters remained in the hands of NISA and the NSC—the same two agencies that had said Japan's reactors were failure proof. If restoring public trust was the government's top priority, many asked, what was the rush? Why not await creation of the new safety agency and give it the authority to develop its own criteria for restarts?

Adding to worries was the release of new seismic data showing that a large portion of western Japan, including Tokyo, was vulnerable to earthquakes nearly twice as large as what the government had projected just nine years earlier. Coastal areas along this same part of Japan also were found to be vulnerable to tsunamis of thirty-two feet (ten meters) or more, three times higher than previous projections. The area around the Hamaoka nuclear plant, southwest of Tokyo, could be hit by a tsunami of sixty-eight feet (twenty-one meters), far above previous predictions. Unlike remote Fukushima, this region was densely populated.

The data also identified previously unknown and potentially active faults located in areas near or even directly underneath, Japanese reactors, including Ohi. Even when presented with the new seismic information, the utilities and government dismissed the potential hazards as exaggerated, without documenting their claims.

"[I]s it really possible to ensure the safety of operating these reactors while the possibility of active faults lying beneath them cannot be ruled out?" the *Asahi Shimbun* asked in an editorial. "The government owes the public a clear and convincing answer to this question."

• • •

The Ohi nuclear plant sits along Japan's scenic west coast, about sixty miles northeast of Osaka, Japan's third-largest city. Comprising four pressurized water reactors, the plant is owned by Kansai Electric Power Company (KEPCO), which relies on nuclear power for almost half its generating capacity, more than any other Japanese utility. Ohi Units 1 and 2 are older PWRs with ice-condenser containments that are smaller and weaker than the large, dry containments at Units 3 and 4. Perhaps for that reason, KEPCO proposed restarting only Units 3 and 4 at first.

In many respects, Ohi served as an ideal test case for both those promoting and those opposing the restarts. The plant is in Fukui Prefecture, home to a total of thirteen reactors. The local economy was heavily dependent on the jobs, tax revenues, and generous perks provided to local communities by utilities doing business there. Some of Japan's largest manufacturers, including Sharp Corporation and Panasonic, are KEPCO customers. The possibility that they and others might move operations offshore if uncertainty regarding energy supplies continued was a clear concern in Tokyo.

Despite their economic benefits, not all neighboring communities were eager to see the Ohi reactors returned so quickly to service. Having a nuclear neighbor once seemed attractive to nearby towns, but that was before they had witnessed the consequences of an accident—the depopulation of a large region, widespread contamination, health risks, uncertainty about a return to normal life. At the height of the restart debate, the government announced that 18 percent of all Fukushima evacuees—including those as far as thirty miles away—might not be able to return to their homes for ten years because of radiation levels.

As a result, some argued that if communities tens or even hundreds of miles away were vulnerable to nuclear plant accidents, they too should have a say in the plants' operation. And the calculus might be different for such communities, since they would be exposed to the risks without receiving any of the benefits. That was something new in Japan.

One of the most outspoken restart critics was Toru Hashimoto, the brash young mayor of Osaka, whose city was Ohi's largest customer and also the largest shareholder in KEPCO. Hashimoto argued that the government was prematurely pushing the restarts at the expense of public safety, a contention that catapulted him into the national spotlight and earned him a large following. Dissatisfied with Tokyo's assurances that the plant was safe, Hashimoto set up his own panel of experts, who took issue with the adequacy of the stress tests. That prompted governors of two nearby prefectures to demand a more thorough inquiry into Ohi's safety before Units 3 and 4 were returned to service.

Critics noted that the stress tests represented just the first phase of what was supposed to be a multipart safety response to the Fukushima accident. Additional measures—construction of higher seawalls, better protection against earthquake damage, improved emergency planning, filtered venting systems—could take years to complete. In the meantime, however, the plants would be operating.

With opposition from local officials and weekly demonstrations in the capital threatening to derail plans for a quick restart at Ohi, Prime Minister Noda stepped up the rhetoric. "Japanese society cannot survive" without the reactors, he warned on national television. And he offered assurances that the government would "continue making uninterrupted efforts" to improve nuclear safety.

The growing pressure from Tokyo and Japan's business community had its effect. Local government opposition to the Ohi restart faded, with officials, including Hashimoto, agreeing to a "limited" restart at least through the summer months.

On June 16, the Noda government officially gave the go-ahead to restart Ohi Units 3 and 4. Trade Minister Yukio Edano acknowledged public opinion polls were running solidly against the restart. "We understand that we have not obtained all of the nation's understanding."

That was evident six days later, when an estimated 45,000 people protested the decision outside the prime minister's office; thousands of demonstrators also rallied in Osaka and elsewhere. The following Friday, an even larger protest massed in the capital near the prime minister's office. As he left for the day, Noda shrugged off the protesters. "They're making lots of noise," he told reporters before heading home.

It would take several weeks for the two Ohi reactors to be brought fully into service, in time to meet what the government had predicted would be a 15 percent shortfall in energy supplies in western Japan during the hottest days of summer. On July 16, with temperatures above ninety degrees, tens of thousands of people clogged central Tokyo protesting the restart. Whether the giant demonstration would have any effect on Japanese energy policy was uncertain. But the crowd, waving banners and holding up signs, made clear that it was time that the public had a say in those decisions. As a marcher told a reporter, "It's just a big step forward to start raising our voices." Despite the protests, in September the government authorized their continued operation.

• • •

In the United States, those who had hoped for change also were feeling frustrated. Among them was the NRC's Gregory Jaczko.

Shortly before leaving his job as chairman in July 2012, Jaczko offered a candid observation to a Washington audience of energy insiders: "I used to say the one thing that kept me up at night was the thing we hadn't thought of. Today, the things that keep me up at night are those things we know we haven't addressed."

It had been a year since the NRC's Near-Term Task Force had made its recommendations on a plan of action, including far-reaching structural reforms. But the slow pace of actual reform bothered Jaczko.

Although many details of the Fukushima Daiichi accident were still being analyzed around the world, its most obvious lessons were clear. Nevertheless, the NRC, bolstered by the task force's statement that the U.S. nuclear fleet did not pose an "imminent risk" to public health and safety, ultimately elected to water down the most pressing recommendations and to slow-walk the others. Just as in the aftermath of Three Mile Island, it appeared that the NRC might be pulling its punches in the face of industry resistance.

In a July 2011 letter to the commission, Marvin Fertel, president of the Nuclear Energy Institute, had thrown down the gauntlet. While conceding that "the industry agrees with many of the issues identified by the task force," he cautioned that "the task force report lacks the rigorous analysis of issues that traditionally accompanies regulatory requirements proposed by the NRC. Better information from Japan and more robust analysis is necessary to ensure the effectiveness of actions taken by the NRC and avoid unintended consequences at America's nuclear energy facilities." The industry appeared to be stalling—demanding further study in the belief that the passage of time would erode momentum for regulatory changes. It had certainly worked in the past.

The task force report also needed to pass muster with the rest of the NRC staff. Although the task force members were NRC personnel, they were operating as an independent body outside of the usual commission pecking order. Normally, safety rule changes evolved slowly within the agency, working their way up to the staff's gatekeeper, the executive director for operations, a post held by Bill Borchardt.

But Chairman Jaczko wanted the commission to vote directly within ninety days on the task force's twelve recommendations, rather than on the staff's views of the report as filtered through Borchardt. Jaczko's attempt to bypass senior management likely grew out of the increasingly rocky relations he was having at the NRC, not only with staff but also with his fellow commissioners, who could join to outvote him. (Jaczko would eventually accuse the majority of favoring policies that "loosened the agency's safety standards.") For their part, the other commissioners expected to also receive the staff's opinions of the report, perhaps broadening their understanding or offering alternatives.

Borchardt and his deputy, Marty Virgilio, believed that the commissioners should receive a "wide range of stakeholder input" before making any decisions on the task force's recommendations. They attached a five-page memo to the task force report making this suggestion. The memo also emphasized that U.S. nuclear power plants were unlikely to experience the same problems as Fukushima had.

Jaczko reportedly disagreed with the stakeholder recommendation and believed that Borchardt's attachment was improperly presented based on procedural grounds. The chairman unilaterally intervened to remove the Borchardt attachment before the task force report was formally transmitted to the commissioners and to Congress. That action, among others, contributed to the internecine conflict that further isolated Jaczko within the agency and ultimately led to his resignation in June 2012.

Although back in July 2011 Jaczko had won the battle over the initial transmittal of the task force report, in subsequent votes he lost the war. Rather than vote directly on the task force's recommendations, the majority ordered the staff to first produce a series of papers analyzing the recommendations and ranking them in priority. But the commissioners made one decision immediately: the task force's highest priority should become the NRC's lowest priority. Recommendation 1, to revise the regulatory framework, was put at the bottom of the list, and the staff was given eighteen months to come up with a proposal for it. In the interim, the commissioners said, the staff should use the NRC's existing regulatory processes to assess the other recommendations.

From one point of view, this move made sense: it could take years to impose safety fixes if they had to await a new regulatory framework. But on the other hand, imposing new requirements under the old framework would only add to the patchwork of regulations and exacerbate the inconsistencies identified by the task force as part of the problem.

Given that the majority of the commissioners did not believe urgent safety upgrades were needed, the necessity for speed was not their motivation for bumping Recommendation 1 off the priority list. Rather, it was driven by the shared view, expressed in a vote by Commissioner William Ostendorff, that the current system was not "broken."[6]

After putting Recommendation 1 aside, the staff proceeded to rank the others in terms of priority by putting them into three tiers. Reevaluating seismic and flood hazards, addressing station blackout risks, and improving containment vents were among the Tier 1 actions that the staff believed "should be started without unnecessary delay."

However, the decision to proceed without first fixing the regulatory framework meant that the new proposals had to navigate all the old impediments to improving safety. In particular, the backfit rule hung over the task force recommendations like the sword of Damocles. Would the NRC buck tradition and acknowledge that U.S. nuclear plants were not adequately protected in light of Fukushima? If not, then most of the changes recommended by the NTTF would be considered "backfits" and would have to meet the NRC's convoluted "substantial safety enhancement" and cost-benefit tests. In that case, there was a good chance that none of the recommendations would ever become a requirement.

But the old system threw up even more hurdles for change. Although Fukushima had proven that beyond-design-basis accidents were a real threat, the NRC's obsolete guidelines still ranked them as very low-probability events, meaning that the calculated benefits of reducing the risk would also be very small. And the benefit could even be zero if the calculation addressed an event that had been left out of the risk assessment models used for the cost-benefit analyses.[7]

The NRC staff ultimately sided with the task force and recommended the commission approve the safety upgrades on the basis of "assuring or redefining the level of protection . . . that should be regarded as adequate." But the majority of the commissioners needed more convincing. Early on, Commissioner Kristine Svinicki warned members of the task force that they were intruding into areas where they could quickly find themselves "swimming in the waters of backfit." In an October 2011 vote, Commissioner Ostendorff asserted that "decisions on adequate protection are among the most significant policy decisions entrusted to the Commission and are not impulsive 'go' or 'no-go' choices." This view, endorsed by the commission majority, led to yet more delay.

However, in a March 2012 decision timed for the first anniversary of Fukushima, all five commissioners approved three orders imposing new regulatory requirements without the need to pass backfit tests. Although the outcome was unanimous, each commissioner had gotten to the destination via a different circuitous route. Most contended that their actions were not expanding adequate protection but merely "ensuring" it. It all added up to a lot of confusion. But in the end, except perhaps for courts of the future that may be called on to parse these distinctions, it didn't really make a difference.

What did matter was how comprehensive and stringent the orders were in addressing Fukushima's lessons. And in those respects, the orders fell short on specifics.

Part of the problem was that the commissioners had directed that the requirements be "performance based." The concept behind performance-based regulation is that requirements should not be too prescriptive. The regulator should specify the desired outcome and let the plant owner figure out the best way to achieve it. While advocates of this approach consider it more efficient because it gives owners more flexibility, it actually makes requirements harder to interpret and enforce. The industry is free to write its own playbook, leaving regulators with the burden of figuring out whether or not it meets the regulatory intent of the safety rules.

As a measure of how popular this approach is with the nuclear industry, over the past couple of decades the NEI has actually drafted guidance documents on complying with performance-based regulations, which the NRC, after some negotiation, then approved.

The B.5.b measures, introduced after 9/11 to help workers cope with the aftermath of an aircraft attack, were one case in which performance-based requirements hadn't worked very well. The industry argued against imposing measures it viewed as too prescriptive or specific, contending that there were simply too many potential disaster scenarios to contemplate. Through their lobbying group, plant owners insisted that they needed maximum flexibility to come up with their own solutions. The NRC relented, imposing only very general requirements for the B.5.b equipment and giving the industry a great deal of leeway to design its own strategies.

However, the industry had not thought through its plans to ensure that they would actually work under real-world conditions, such as high radiation fields, excessive heat, and infrastructure damage. These were the very conditions that contributed to the failure of the Japanese severe accident management measures at Fukushima. Yet in a vote on the Fukushima proposals, the commissioners endorsed the B.5.b process as a model for dealing with beyond-design-basis accidents.

Two of the NRC's new orders were relatively uncontroversial. One required the installation of "reliable hardened vents" at Mark I and Mark II boiling water reactors so that they could be used under station blackout conditions. Previously plant owners had been allowed to install these vents voluntarily, so that NRC inspectors had no power to review their adequacy or require improvements. This situation, a legacy of the NRC's timid approach to dealing with severe accidents in the 1980s, provides little confidence that vents at U.S. BWRs would have worked any more effectively than those at Fukushima. The second order called for reliable instrumentation in spent fuel pools, something that could have helped workers at Fukushima better understand the unfolding situation.[8]

The third order directed plants to develop "mitigation strategies" for beyond-design-basis external events. But the title promised more than the order actually delivered. Plants were required to be "capable of mitigating a simultaneous loss of all alternating current (AC) power and loss of normal access to the ultimate heat sink" for all units at a site. As the task force had recommended, the strategies would involve three phases: using installed equipment, using on-site portable equipment, and using off-site equipment. But where the task force had recommended that specific durations be set for the first two phases—namely, installed equipment should be able to maintain cooling for eight hours and portable equipment for the next seventy-two hours—the order did not specify minimum durations for these phases. Each plant site would propose its own.[9]

IT'S RISKY WHEN INDUSTRY WRITES THE RULES

The NRC's acquiescence to the FLEX program of rapidly deployable emergency equipment was typical of the way the agency and the industry's advocacy group, the NEI, have interacted for many years in interpreting regulations and developing compliance documents.

In the early days, when a regulation or order needed interpreting, the NRC would develop a "regulatory guide" and submit it for comment by the affected industry. But over time the roles were often reversed: the industry, coordinated by NEI, would write the first drafts of guidance documents and the NRC would comment. This shift gave the industry far greater power to shape the conceptual basis for regulatory compliance, leaving the NRC to tinker around the edges. The NEI guidance documents would then serve as boilerplate text for use by all applicants and licensees.

The NEI guidance for the FLEX program was such a template. The NRC required all plant owners to submit plans by the end of February 2013 to demonstrate how they would implement the program to comply with the mitigation strategies order. On February 28, Exelon Corporation, which owns seventeen reactors, submitted its plan for the Peach Bottom plant in south central Pennsylvania. Exelon's proposal was typical of all the plant owners' responses, closely hewing to the NEI's guidance. It clearly demonstrated the shortcomings of the FLEX approach.

The Peach Bottom plant, situated beside the Susquehanna River, has two Mark I BWRs closely resembling Fukushima Daiichi Units 2 and 3. Exelon's plan for dealing with a beyond-design-basis accident there assumed from the get-go that batteries and electrical distribution systems for both AC and DC power would be available. It assumed that off-site personnel called to duty would be able to reach the plant in as little as six hours, not necessarily a realistic assumption in the event of the type of major natural disaster that the emergency plans were being designed for.

The Peach Bottom Atomic Power Station, located on the Susquehanna River in south-central Pennsylvania, has two Mark I boiling water reactors, similar to Units 2 and 3 at Fukushima Daiichi. Emergency response plans for the plant depend upon many assumptions now proven to be unrealistic in the aftermath of events at Fukushima Daiichi. *U.S. Nuclear Regulatory Commission*

The plan also set time frames for coping with an extended station blackout that fell short of the periods recommended by the NTTF. The task force wanted plants to be able to keep fuel cool for the first eight hours of a blackout using permanently installed equipment and thereafter using portable equipment (such as FLEX items) that was available on-site for the next seventy-two hours. After that, one could assume that additional equipment and supplies delivered from off-site would be available.

For Peach Bottom, Exelon estimated that the batteries needed to run critical equipment, including the RCIC, would last no longer than five and a half hours—but it also asserted that the FLEX generator could be hooked up and ready to start recharging the batteries within five hours. Therefore, there was no need for Exelon to extend the battery capacity.

Exelon's plan assumed that additional backup equipment from the closer of two Regional Response Centers, located nearly one thousand miles away in Memphis, would arrive at Peach Bottom within twenty-four hours. (Exelon contended that it could keep the plant stable forever without any off-site assistance, with repeated torus venting, but conceded that it would be better to eventually have an alternative, less radioactive method.)

Consistent with NEI's guidance, Exelon's plan allowed FLEX equipment to be stored below flood level on the assumption that workers would have time to move it to a safer place "prior to the arrival of potentially damaging flood levels."

As questionable as these timelines are, they don't take into account the worst case. Exelon found that if a station blackout occurred during refueling, when the entire core of a unit was in the spent fuel pool, there would be no time for the first coping phase at all: FLEX equipment would have to be set up immediately. Exelon asserted that this was not a problem because even though the pool would start to boil after two and a half hours, the spent fuel would not become uncovered until eight hours had elapsed.

The Peach Bottom FLEX plan is a perfect example of the mind-set that led to Fukushima. It represents industry and regulators scripting an accident with little room for improvisation. If a single assumption fails—say, that workers don't have time to move FLEX equipment to safety in advance of an impending flood—then all the other barriers would collapse like dominoes. Without the FLEX generator, the batteries would fail after five and a half hours and the RCIC could no longer be counted on to cool the reactors. Operators would eventually lose the ability to vent the containment and it would over-pressurize. Backup equipment, located a thousand miles away in Tennessee—and possibly on the other side of massive floodwaters or earthquake destruction—probably would not arrive in time to save the day. Nobody apparently thought those possibilities were worth considering, even after Fukushima Daiichi.

The order also specified that reactor owners must provide "reasonable protection" of the emergency equipment from external events. But what did that mean? The definition was apparently in the eye of the beholder. For instance, the task force had said the equipment should be protected against beyond-design-basis floods, but the commission's order contained no such requirement.

In any event, by the time the NRC issued the mitigation-strategies order on March 12, 2012, it no longer seemed to matter much. While the NRC commissioners had been ruminating about adequate protection, the industry was creating its own facts on the ground. Plant sites had already ordered or acquired more than three hundred pieces of major equipment under the FLEX voluntary initiative. "Time is of the essence," said Charles "Chip" Pardee, then head of power generation at Exelon. Arguing that the industry needed to move forward to meet the NRC's timeline for safety improvements that were still being formulated, Pardee said, "We are proceeding informed by what the NRC is doing, but [we are] out in front of regulations."

The FLEX approach, however, arguably did not go nearly as far as was needed. For instance, the industry insisted that its purpose was to prevent core damage. Thus, the program did not contemplate how to use such equipment in the highly radioactive and hot environment that would occur after a core started to melt—even though that was one of the greatest challenges at Fukushima. And FLEX was designed with the assumption that a plant's electrical distribution systems were functional, also at odds with the reality experienced at Fukushima.

The industry's eagerness to address the lessons of Fukushima even before the NRC could issue requirements was highly ironic at best. After all, plant owners at the same time were dragging their feet—sometimes for decades—on addressing any number of outstanding safety problems, from fire protection to emergency core cooling system blockages. Getting "out in front of regulations" with FLEX was a brilliant move. The more money the industry was spending on equipment that met its own specifications, the more difficult—politically and practically—it would be for the NRC to require substantially different and more robust measures.

The tactic worked. The FLEX program, for all its flaws, did not conflict with the NRC's ambiguous mitigation strategies order. After many months of deliberation with the NEI on the guidance document it had prepared for the use of FLEX equipment, the NRC largely endorsed NEI's approach. The tail had wagged the dog.

Over the summer of 2012, the Japanese government solicited views from citizens as it drafted a new energy policy. The results were decisive, noted the survey organizers: "We can say with certainty that a majority of citizens want to achieve a society that does not rely on nuclear power generation."

Although some believed that the government's outreach was more appearance than substance, there were signs that the public's dissatisfaction with the Japanese establishment might be having an impact elsewhere. In September TEPCO announced the creation of an outside review committee to oversee its nuclear operations. With guidance from the committee, TEPCO's new management hoped to win approval to restart the seven reactors at Kashiwazaki-Kariwa beginning in the spring of 2013, apparently confident that the government's effort to craft a new energy policy curtailing nuclear power would go nowhere.

On September 14, 2012, the Noda government unveiled its new energy plan, the keystone of which was the elimination of nuclear dependency by the end of the 2030s. The plan was short on specifics. As the *Financial Times* noted, it was "a somewhat messy compromise that will delight nobody." It definitely did not delight Japan's business leaders, who lobbied the Noda government to retain the status quo. "It is highly regrettable that our argument was comprehensively dismissed," said the head of the powerful business group Keidanren.[10]

Noda warned that moving toward a nuclear-free Japan would not be easy. "No matter how difficult it is," he said, "we can no longer put it off."

It took all of a week before the energy plan was shot full of so many holes that it was barely recognizable.

Noda, like Kan before him, was struggling for political survival. His popularity ratings had not fallen quite as far as Kan's had a year earlier, but Noda's party, the Democratic Party of Japan, had an approval rating in the teens. New elections, expected soon, almost certainly would sweep another party and prime minister into office.

Noda's energy plan originally called for shutting down reactors at the end of their forty-year operating lives and building no new ones. But the day after its release, Yukio Edano, the trade minister, said the ban on new construction would not apply to three reactors already approved for construction. Even more illogically, Japan would continue to pursue its ambitious plan to begin full-scale operation of the Rokkasho Reprocessing Plant, a facility for reprocessing spent nuclear fuel, even though presumably there would be little reactor capacity to dispose of the many tons of plutonium it would produce each year. Operating Rokkasho would increase the massive stockpile of weapon-usable plutonium that Japan had already accumulated, a major security and safety concern.

It seemed that Noda, given the choice of appeasing Japan's powerful business community or all those people "making lots of noise" on the streets, had opted for the establishment. The backpedaling continued: when the cabinet met on September 19, it sidestepped approving the energy plan, saying only it would "take it into consideration." The cabinet's failure to endorse the plan meant that future governments would not be required to follow it. That prompted the *Asahi Shimbun* to note, "The government's commitment to abandon nuclear power by the 2030s is increasingly sounding like 'maybe.'"

The cabinet did approve creation of a new nuclear watchdog, the Nuclear Regulation Authority (NRA), after months of political wrangling over the extent of its independence. Even that was cast in doubt when it was reported that most NRA employees had come from the widely discredited NISA and the NSC. The cabinet handed the new agency the unenviable task of deciding whether the nation's nuclear fleet could safely be restarted. The NRA said the decision could take months. The agency's rocky start only signaled the tough job it faced in ushering in a new era of nuclear oversight.

The media reported that four of six experts charged with drawing up new safety regulations for the NRA had financial ties to the nuclear industry. (Trying to put the best face on that news, an NRA spokesman noted that the mere fact that the experts had to disclose their personal finances was a positive step.)

In January 2013, the NRA issued a draft set of new safety measures that nuclear plants would have to adopt as a prerequisite for restart. These included upgrades to some of the equipment that had proved inadequate at Fukushima, such as portable power supplies and water sources, so that they would be more reliable in an extended station blackout or other severe accident. Plants would have to show that they could maintain core cooling for a full week without off-site assistance. They would also have to install filtered vents and remote secondary control rooms. In addition, they would have to demonstrate defense against both natural disasters and terrorist attacks, such as a 9/11-style aircraft impact.

The NRA then began a period of "discussion" with the utilities to obtain their views before finalizing the new rules. This was uncharted territory in Japan. It remained to be seen if the NRA would be able to maintain its independence when confronted with predictably intense pressure from the utilities to weaken this stringent and costly set of proposals.

Even as regulators in Japan's nuclear establishment haggled over details, those whose lives had been disrupted by events at Fukushima Daiichi continued to struggle. For many of them, resuming their normal lives anytime soon seemed unlikely.

A case in point were residents of the "emergency evacuation preparation" zone—a portion of the area between twelve and eighteen miles around the plant. About a month after the accident, the government recommended that vulnerable groups such as children and pregnant women in this zone evacuate, and that all others be ready to flee at a moment's notice. (In contrast, the highly contaminated area to the northwest of the plant, extending beyond eighteen miles, was a mandatory evacuation zone and remained so.) Nearly half of the 58,000 people in the "evacuation preparation" zone left their homes.

When protective measures were lifted six months later, the evacuees stayed away. Only about 3,100 people had returned as of September 2012. Noticeably absent were young children.

Restricted Area, Deliberate Evacuation Area, and Evacuation-Prepared Area in Case of Emergency and Regions, Including Specific Spots Recommended for Evacuation (as of August 3, 2011)

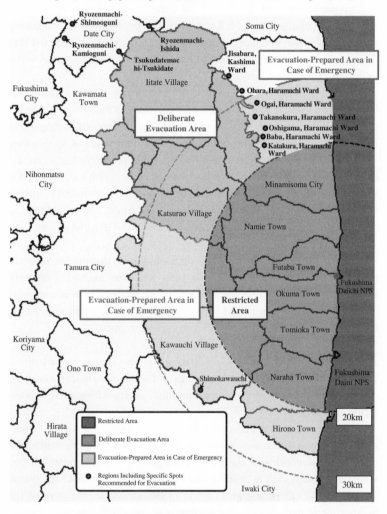

More than a year after the accident, large areas near Fukushima Daiichi remained off-limits. *Investigation Committee on the Accident at the Fukushima Nuclear Power Stations of Tokyo Electric Power Company*

For residents of the town of Okuma, home to Fukushima Daiichi, the announcement that it would not be safe to return until 2022 was the first blow delivered by the government. Then came word that Tokyo planned to build as many as nine temporary radioactive disposal facilities in Okuma—possible because no one was living there. That spawned fears that even when it was safe to go back, no one would want to. "We have been living there for 1,000 years," Okuma's mayor told a reporter.

In October 2012, some of the workers who had risked their lives to wrestle Fukushima Daiichi back from the brink spoke publicly about their experiences for the first time. Until that moment these men had avoided the limelight. Eight of them met with Prime Minister Noda, who was visiting the plant. They began with an apology to the Japanese people.

Six of the workers declined to be identified, or to have their faces shown on camera. "Many [people] view us as the perpetrators," explained Atsufumi Yoshizawa, who worked with plant superintendent Masao Yoshida during the accident. Uppermost in their minds as the accident unfolded, the men said, was the safety of their fellow workers and of nearby communities, where many of them lived with their families.

They said they had never intended to abandon the plant during the crisis; they knew that erroneous reports to that effect had led to an angry exchange between Prime Minister Kan and TEPCO's president. "I thought that maybe I would end up not leaving," Yoshizawa said, "that, as we Japanese say, we would 'bury our bones' in that place."

But the workers stayed despite the terrifying conditions. "I had no intention of dying," said the operations chief for Units 1 through 4. "Everyone did their best. Dying would have meant giving up."

Noda returned to Fukushima Prefecture in early December, to kick off the campaign season for the December 16 parliamentary election. Also in Fukushima that day was Shinzo Abe, seeking reelection to the prime minister's job that he resigned in 2007 after a troubled year in office. For voters across Japan, the slumping economy occupied center stage, and Abe promised change. When the votes were counted, Abe and his conservative-leaning Liberal Democratic Party won by a landslide.

By month's end, Abe was sworn in as Japan's seventh prime minister in six and a half years. Among his first promises was a vow to move ahead with new nuclear development.

12

A RAPIDLY CLOSING WINDOW
OF OPPORTUNITY

The accusations began flying almost immediately. Over the ensuing weeks and months, everyone, it seemed, was looking for something or someone to blame for the disaster at Fukushima Daiichi. Antinuclear activists around the globe looked at the accident and saw irredeemable technical and institutional failures, reinforcing their conviction that nuclear energy is inherently unsafe. Nuclear power supporters indicted the user of the technology—in this instance TEPCO—instead of the technology itself, and thus avoided answering larger safety questions.

TEPCO initially pointed fingers at Mother Nature, asserting that the event was unavoidable. But the utility also blamed Japanese government regulators for not forcing the company to meet sufficiently stringent safety standards.

For its part, the Japanese government blamed the "unprecedented" accident on what it called "an extremely massive earthquake and tsunami rarely seen in history," suggesting that there was no way authorities could have anticipated such an event. By invoking forces beyond their control, TEPCO and the Japanese government absolved themselves of responsibility.

Some experts in the United States and Japan decided the culprit was the deficient Mark I reactor design, contending that advanced reactor designs, such as the Westinghouse AP1000, would not have suffered the same fate. In response, the Mark I's designer, GE, asserted that no nuclear plant design could have survived the flooding and blackout conditions that Fukushima Daiichi experienced.

Then there were those who argued precisely the opposite: that the Fukushima disaster was clearly preventable. The Japanese Diet Independent Investigation Commission laid blame at the feet of both TEPCO and Japanese regulators for their incompetence, lack of foresight, and even corruption.

The utility and government should have anticipated and prepared for larger earthquakes and tsunamis, and should have had more robust accident management measures in place, the commission concluded.

It is too simplistic to say either that the accident was fully preventable or that it was impossible to foresee. The truth lies in between, and there is plenty of blame to go around.

NEW REACTOR DESIGNS: SAFER OR MORE OF THE SAME?

The accident at Fukushima provided nuclear advocates with a fresh argument to bolster their support for construction of a new generation of reactors. Certain new designs, they claimed, could have withstood a catastrophic event such as the disaster at Fukushima Daiichi.

Among those designs is the AP1000, built by Westinghouse Electric Company (AP stands for "advanced passive" and one thousand is its approximate electrical generation output in megawatts.) The reactor utilizes passive safety features to reduce the need for machinery, such as motor-driven pumps, to provide coolant in the event of an accident. Westinghouse advertises the AP1000 as capable of withstanding a seventy-two-hour station blackout. Soon after Fukushima, Aris Candris, then CEO of Westinghouse, said that the Fukushima accident "would not have happened" had an AP1000 been on the site.

The design received a significant boost in February 2012 when the NRC commissioners voted 4–1 to approve licenses to construct two AP1000 reactors at the Vogtle nuclear complex, about 170 miles east of Atlanta. NRC chairman Gregory Jaczko cast the lone "no" vote, saying, "I cannot support issuing this license as if Fukushima had never happened." Jaczko anticipated that the lessons learned from the Japanese disaster would lead the NRC to adopt new requirements for U.S. plants, and he argued that any new licenses should be conditioned on plant owners' agreement to incorporate all safety upgrades that the NRC might require in the future.

Other new reactor designs, such as the GE Hitachi ESBWR ("economic simplified boiling water reactor"), also boast passive systems. In contrast, the French company Areva's EPR ("evolutionary power reactor" in the United States, "European pressurized water reactor" elsewhere) adds redundant active safety systems and new features to cope with severe accidents such as a "core catcher," intended to capture and contain a damaged core if it melts through the reactor vessel. Both designs await NRC certification in the United States.

Advocates are also promoting development of small modular reactors, those generating less than three hundred megawatts of electricity. In principle, the units can be built on assembly lines and would allow utilities to augment their generating output in smaller increments to meet fluctuating demand.

Backers argue that because the small units could be constructed underground, they would be safer from terrorist attacks or natural events such as earthquakes. On the other hand, critics argue that in the case of a serious accident, emergency crews would have difficulty accessing the below-grade reactors, and multiple units at one site may compound the challenges for emergency response efforts, as was seen at Fukushima.

There is no question that nuclear safety can be improved through thoughtful design of new reactors. However, nuclear power's safety problems cannot be solved through good design alone. Any reactor, regardless of design, is only as robust as the standards it is required to meet.

Unless regulators expand the spectrum of accidents that plants are designed to withstand, even enhanced safety systems could prove of little value in the face of Fukushima-scale events such as an extended station blackout or a massive earthquake.

In the case of the AP1000, for example, Aris Candris was wrong. Even if an AP1000 had been at the Fukushima site, it would have become endangered after three days of blackout when it would have needed AC power to refill the overhead tank that supplies the emergency cooling water. And if the plant had experienced an earthquake larger than it was designed to withstand, the tank itself might have been rendered unusable.

Perhaps the strongest vote of no confidence comes from the reactor vendors themselves. Even as they heavily promoted the new designs in the United States, the vendors in 2003 successfully lobbied Congress to reauthorize federal liability protection for all reactors—new and old—under the Price-Anderson Act for another twenty years. While they asserted that the next generation of plants would pose an infinitesimally small risk to the public, they wanted to make sure there would be limits on the damage claims they would have to pay if they were wrong.

Those with an interest in advancing nuclear safety need to look past their own biases and recognize the root causes of the accident, wherever they may be. Absent such a clear-eyed appraisal, the opportunity to identify and correct past mistakes will be lost. With it will go the opportunity to prevent the next severe accident. The legacy of Fukushima Daiichi will then be revealed as a tragic fiasco.

There's no question that TEPCO and the Japanese regulatory system bear much responsibility. Each clearly could have done more to prevent the disaster (like erecting a taller seawall when information about a larger tsunami threat emerged) or to lessen its severity (like equipping the plant with more reliable containment vent valves). Such change could have been accomplished through new regulations or voluntary initiatives, neither of which was forthcoming.

But that is too narrow a focus. TEPCO and government regulators were merely the Japanese affiliate of a global nuclear establishment of power companies, vendors, regulators, and supporters, all of whom share the complacent attitude that made an accident like Fukushima possible.

The safety philosophy and regulatory process that governed Fukushima were not fundamentally different from those that exist elsewhere, including the United States. The reactor technology was nearly identical. The reality is that any nuclear plant facing conditions as far beyond its design basis as those at Fukushima would be likely to suffer an equivalent fate. The story line would differ, but the outcome would be much the same—wrecked reactors, off-site radioactive contamination, social disruption, and massive economic cost.

The catastrophe at Fukushima should not have been a surprise to anyone familiar with the vulnerabilities of today's global reactor fleet. If those vulnerabilities are not addressed, the next accident won't be a surprise, either.

Fukushima triggered extensive "lessons learned" reviews in Japan, France, the United States, and elsewhere. Many lessons have indeed been learned, but to date few have been promptly and adequately addressed—at least in the United States. The reason, of course, is the prevailing mind-set.

"It Can't Happen Here"

That mind-set underlies all that went wrong at Fukushima Daiichi. While numerous technological and regulatory failures contributed to the events that began on March 11, 2011, that pervasive belief stands as the root cause of the accident.

In the United States, "it can't happen here" was a common refrain while details of the Fukushima accident were still unfolding. In June 2011, for example, Senator Al Franken joked that "the chances of an earthquake of that level in Minnesota are very low, but if we had a tsunami in Minnesota, we'd have bigger problems than even the reactor."

Senator Franken's casual attitude illustrates the problem. Yes, it is unlikely that a tsunami will sweep into the northern plains. But serious potential threats to reactors do exist in his home state, as well as the states of many other members of Congress.

Two pressurized-water reactors at the Prairie Island nuclear plant, southeast of Minneapolis, are among the thirty-four reactors at twenty sites around the United States downstream from large dams. A dam failure could rapidly inundate a nuclear plant, disabling its power supplies and cooling systems, not unlike the impact of a tsunami.[1]

Nor is a dam failure the only type of accident that could create Fukushima-scale challenges at a U.S. nuclear plant. Fire is another. A fire could damage electric cabling and circuit boards, cutting off electricity from multiple backup safety systems as flooding did at Fukushima. The NRC adopted fire-protection regulations in 1980 following a very serious fire in March 1975 at the Browns Ferry nuclear plant in Alabama. A worker using a lit candle to check for air leaks accidentally started a fire in a space below the control room. The fire damaged electrical cables that disabled all of the emergency core cooling systems for the Unit 1 reactor and most of those systems for the Unit 2 reactor. Only heroic actions by workers prevented dual meltdowns that day at Browns Ferry.

The threat of fires remains a major contributor to the risk of core damage at nuclear plants. Decades later, fire safety regulations imposed by the NRC in the wake of Browns Ferry have not been met at roughly half the reactors operating in the United States—including the three reactors at Browns Ferry.

So, lesson one: "It *can* happen here." And that reality leads directly, once again, to the most critical question.

"How Safe Is Safe Enough?"

Fukushima and other nuclear plants are not houses of cards waiting for the first gust of wind or ground tremor to collapse. They are generally robust facilities that require many things to go wrong before disaster occurs. Over the years, many things have, mostly without serious consequences.

Key to safety at a nuclear plant are multiple barriers engineered to protect the public from radiation releases—the so-called defense-in-depth approach. Each barrier should be independent of the others and provide a safety margin exceeding the worst conditions it might have to endure in any accident envisioned in the design basis of the plant. Regulators regard defense-in-depth as a hedge against uncertainty in the performance of any one barrier.

The design and operation of the damaged reactors at Fukushima Daiichi were consistent with the defense-in-depth principle as commonly applied around the world. Multiple and diverse cooling systems existed to forestall damage to the reactor cores. When cooling was interrupted and core damage did occur, leak-tight containment buildings limited the escape of radioactivity.

When containments leaked and a large amount of radioactivity escaped, emergency plans were invoked to evacuate or shelter people.

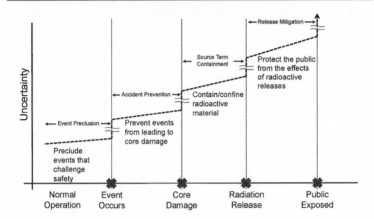

U.S. Nuclear Regulatory Commission

Defense-in-depth works well—as far as it goes. The concept has succeeded in limiting the frequency of nuclear disasters. Since the 1970s, three nuclear plant accidents have drawn international attention: Three Mile Island in 1979, Chernobyl in 1986, and Fukushima in 2011. Were it not for defense-in-depth, the list could be larger.

Hurricane Andrew, which pummeled Florida in 1992, extensively damaged the Turkey Point nuclear plant near Miami, but defense-in-depth prevented disaster. The plant lost access to its off-site electric power grid for several days, but emergency diesel generators powered essential equipment in the meantime. And while winds knocked over a water tower and extensively damaged a warehouse and training building on site, more robust structures protected other water supplies and essential equipment.

A month after Fukushima, a severe tornado disconnected the Browns Ferry plant from the electrical grid. That June, the Fort Calhoun nuclear plant in Nebraska experienced flooding that temporarily made it an island in the Missouri River, and in August the North Anna nuclear plant in Virginia was shaken by an earthquake of larger magnitude than it was designed to withstand. In each case, the conditions caused by the accident did not exceed the safety margins to failure—from a loss of off-site power at Browns Ferry, a flood at Fort Calhoun, or an earthquake at North Anna.

But did these events prove the inherent safety of nuclear plants, as the U.S. industry and regulators claim? Or did they constitute accidents avoided not by good foresight, but rather by good fortune?

For all its virtues, defense-in-depth has an Achilles' heel, one rarely mentioned in safety pep talks. It is known as the common-mode failure. That happens when a single event results in conditions exceeding the safety margins of *all* the defense-in-depth barriers, cutting through them like a hatchet through a layer cake.

Common-mode failure is what flooding caused at Fukushima and what fire caused at Browns Ferry. Flooding or fire took out all the redundant systems needed to cool the reactor cores, the systems needed to keep the containments from overheating and leaking, and the systems needed to help predict the path and extent of the radioactive plumes. At Browns Ferry, workers managed to employ ad hoc measures in time to prevent disaster. At Fukushima, time ran out.

Defense-in-depth is both a blessing and a curse. It allows many things to go wrong before a nuclear plant disaster occurs. But when too many problems arise or a common-mode failure disables many systems, defense-in-depth can topple like a row of dominoes. The risk of common-mode failure can be reduced through enhancing defense-in-depth, but it can never be eliminated.

The true curse of defense-in-depth is that it has fostered complacency. The existence of multiple layers of defense has excused inattention to weaknesses in each individual layer, increasing the vulnerability to common-mode failure.

Fukushima Daiichi was a well-defended nuclear plant by accepted standards, with robust, redundant layers of protection. When the earthquake knocked out the off-site electrical grid, emergency diesel generators stood ready as the backup power source. Each of the six reactor units had at least two of these generators (one unit had three). A single emergency generator could provide all the power needed for cooling a core and other essential tasks, but defense-in-depth made sure every reactor had at least one spare.

It didn't help. The generators were protected, like the reactors themselves, by the seawall erected along the coast. When the tsunami washed over the seawall, it disabled all but one of the emergency generators or their electrical connections.

Even without the generators, defense-in-depth offered protection. Banks of batteries were ready to power a minimal subset of safety equipment while damage to the AC power systems was being repaired. The battery capacity at Fukushima was eight hours per unit, assumed to be ample time for workers to either restore an emergency diesel generator or recover the electrical grid. But it took nine days to partially reconnect the plant to the grid and even longer to restore the generators.

At Fukushima, as in the United States and elsewhere, reactor operators were trained in emergency procedures for responding to severe accidents. These procedures instructed them to take steps like venting the containment to reduce dangerously high pressure and enable cooling water to enter. But the manuals did not envision the conditions the operators actually faced—for example, the need to operate vents manually in darkness—and thus workers could not implement these procedures in time to prevent the meltdowns.

The last defense-in-depth barrier was evacuation. But at Fukushima, emergency planning proved ineffective at protecting the public. Evacuation areas had to be repeatedly expanded in an ad hoc manner, and in some cases the decisions were made far too late to prevent radiation exposures to many evacuees.

All of Fukushima's defensive barriers failed for the same reason. Each had a limit that provided too little safety margin to avert failure. Had just one barrier remained intact, the plant might well have successfully endured the one-two punch from the earthquake and tsunami or at the bare minimum, the public would have been protected from the worst radiation effects.

The chance that *all* the barriers might fall was never part of the planning. In effect, the nuclear establishment was riding a carousel, confident that the passing scenery of anticipated incidents would never change. Lost in the process was this reality: the brass ring for this not-so-merry-go-round involves both foreseeing hazards *and* developing independent, robust defense-in-depth barriers to accommodate unforeseen hazards. One without the other has been repeatedly shown, at tremendous cost, to be insufficient.

Severe accidents like Fukushima render the status quo untenable for the nuclear establishment. Something has to change. But meaningful change—a true reduction in potential danger to the public—will not occur until regulators and industry look ahead and to the sides, at what *could* happen, not solely at what *has* happened.

But such an approach would be a turnabout for the nuclear establishment. Historically, when responding to events like Three Mile Island, the industry and regulators have worked hard to narrow the scope of the response, simply patching holes in the existing safety net rather than asking whether a better net is needed. To put it bluntly, unless this process is radically overhauled, it will take many more nuclear disasters and many more innocent victims to make the safety net as strong as it should be today.

Highway departments could put up roadside signs saying "Don't Go Too Fast" or "Drive at a Reasonable Pace." Instead, they put up signs reading, for example, "Speed Limit 55" or "Maximum Speed 20" so that drivers understand what is expected and law enforcement officers know when to issue traffic tickets. The former signs would constitute entirely useless measures: a car wrapped around a tree must have been traveling too fast, but another barreling through a school zone at 120 miles per hour must be operating at a reasonable speed if it doesn't strike any children. Safety requires specificity. Lack of specificity invites a free-for-all.

Such is precisely what the NRC's Near-Term Task Force tried to avoid in its proposals for reducing vulnerabilities at U.S. nuclear plants. The NTTF started its report with a recommendation for fundamental change. It called on the NRC to redefine its historical safety threshold of "adequate protection," this time establishing a clear foundation to guide both regulators and plant owners in addressing beyond-design-basis accidents.

In essence, the NTTF's first recommendation urged the NRC to formally recognize that beyond-design-basis accidents need to be guarded against with unambiguous requirements based on robust defense-in-depth *and* well-defined safety margins. In this way, plant owners as well as NRC reviewers and inspectors would better be able to agree on what was acceptable. Inspectors would have a legal basis for declaring violations should plant owners fail to meet the requirements; at the same time, plant owners would be better protected from arbitrary rulings.

The NTTF did not propose taking a sledgehammer to the existing system. Instead it recommended creating a new category of accident scenarios to cover a range of possibilities not envisioned in the design bases of existing nuclear plants. This new category of "extended design-basis accidents" could include aspects of some of the extreme conditions experienced at Fukushima. A new set of regulations would be created for extended design-basis accidents, eliminating the "patchwork" that currently governed beyond-design-basis accidents. However, the requirements themselves would be less stringent than those for design-basis accidents.

The task force believed that the NRC had come to depend too much on the results of highly uncertain risk calculations that reinforced the belief that severe accidents were very unlikely. That, in turn, had provided the NRC with a rationale to shrink safety margins and weaken defense-in-depth. To remedy the problem, the task force requested that the commission formally consider "the completeness and effectiveness of each level of defense-in-depth" as an essential element of adequate protection. The task force also wanted the NRC to reduce its reliance on industry voluntary initiatives, which were largely outside of regulatory control, and instead develop its own "strong program for dealing with the unexpected, including severe accidents."

The task force members believed that once the first proposal was implemented, establishing a well-defined framework for decision making, their other recommendations would fall neatly into place. Absent that implementation, each recommendation would become bogged down as equipment quality specifications, maintenance requirements, and training protocols got hashed out on a case-by-case basis.

But when the majority of the commissioners directed the staff in 2011 to postpone addressing the first recommendation and focus on the remaining recommendations, the game was lost even before the opening kickoff.

The NTTF's Recommendation 1 was akin to the severe accident rulemaking effort scuttled nearly three decades earlier, when the NRC considered expanding the scope of its regulations to address beyond-design accidents. Then, as now, the perceived need for regulatory "discipline," as well as industry opposition to an expansion of the NRC's enforcement powers, limited the scope of reform. The commission seemed to be ignoring a major lesson of Fukushima Daiichi: namely, that the "fighting the last war" approach taken after Three Mile Island was simply not good enough.

Consider the order for mitigation strategies issued by the NRC to all plant owners on March 12, 2012, a year and a day after Fukushima. One part of the order required that plant owners "provide reasonable protection for the associated equipment from external events" like tornadoes, hurricanes, earthquakes, and floods. "Full compliance shall include procedures, guidance, training, and acquisition, staging, or installing of equipment needed for the strategies," the order read.

But what is "reasonable protection"? What kind of "guidance" would be adequate, and how rigorous would be the training be? The NTTF's first proposal would have required specific definitions. Now, without a concrete standard, NRC inspectors will be ill equipped to challenge protection levels they deem unreasonable. Conversely, plant owners will be defenseless against pressure from the NRC to provide more "reasonable" levels of protection.[2]

A second example: another order the NRC issued on March 12, 2012, required the owners of boiling water reactors with Mark I and II containments to install reliable hardened containment vents. They left for another day a decision on whether the radioactive gas from containment should be filtered before being vented to the atmosphere.

Such vents are primarily intended to be used before core damage occurs, when the gas in the containment would not be highly radioactive and filters would not normally be needed. However, as Fukushima made clear, it is possible that the vents will have to be used *after* core damage occurs, to keep it from getting worse. In that case, filters could be crucial to reduce the amount of radioactivity released. The presence of filters could also make venting decisions less stressful for operators in the event that they weren't sure whether or not core damage was taking place. In November 2012, the NRC staff recommended that filters be installed in the vent pipes, primarily as a defense-in-depth measure.

But what might seem a simple, logical decision—install a $15 million filter to reduce the chance of tens of billions of dollars' worth of land contamination as well as harm to the public—got complicated. The nuclear industry launched a campaign to persuade the NRC commissioners that filters weren't necessary. A key part of the industry's argument was that plant owners could reduce radioactive releases more effectively by using FLEX equipment.

Vent filters would only work, the argument went, if the containment remained intact. If the containment failed, radioactive releases would bypass the filters anyway. And sophisticated FLEX cooling strategies could keep radioactivity inside the containment in the first place. Further, the absence of filters at Fukushima might not have caused the widespread land contamination in any event because it wasn't clear that the largest releases occurred through the vents; the radioactivity may have escaped another way.

The NRC staff countered by claiming that the FLEX strategies were too complicated to rely on: they rested on too many assumptions about what might be taking place within a reactor in crisis and what operators would be capable of doing. In contrast, a passive filter could be counted on under most circumstances to do its job—filter any radioactivity that passed through the vent pipes. The staff argued that filters would be warranted as a defense-in-depth measure. The staff also pointed out that many other countries, like Sweden, simply required vent filters to be installed decades ago as a prudent step.

Without an explicit requirement to consider defense-in-depth, as the NTTF had called for in its first recommendation, the NRC commissioners could feel free to reject the staff's arguments. In March 2013, they voted 3–2 to delay a requirement that filters be installed, and recommended that the staff consider other alternatives to prevent the release of radiation during an accident. However, at the same time the commissioners voted to require that the vents themselves be upgraded to be functional in a severe accident. This second decision didn't make much sense: if the commissioners believed the vents might be needed in a severe accident, then what excuse could there be for not equipping them with filters?

As the Fukushima disaster recedes in public memory, and the NRC sends mixed signals, the nuclear establishment has begun relying on the FLEX program to address more and more of the severe accident safety issues identified by the task force and the NRC staff. It has come to see FLEX not as a short-term measure but as an enduring solution—all that is needed to patch any holes in the safety net.

Three months before the vote on the filters, the commissioners decided to slow down the development of a new station blackout rule—one they had originally identified as a priority. In part, it was because of their confidence in FLEX. The NRC later notified Congress that it had rejected a number of other safety proposals, such as adding "multiple and diverse instruments to measure [plant] parameters" and requiring all plants to "install dedicated bunkers with independent power supplies and cooling systems," because it believed that FLEX was sufficient. But it is unclear how FLEX would obviate either the need for additional reliable instrumentation or the desirability of a bunkered emergency cooling system. Indeed, both measures would support FLEX capabilities should a crisis occur.

Despite the willingness of the NRC to give FLEX a resounding vote of confidence, there were three important defense-in-depth issues that could not be dismissed with a wave of the FLEX magic wand. These issues had been flagged by the NRC staff post-Fukushima as issues that were of great public concern and deserved consideration.

The first related to the U.S. practice of densely packing spent fuel pools, a situation that some critics had long pointed to as a safety hazard and that the NRC had long countered was perfectly safe.

In the aftermath of Fukushima, it turned out that the Unit 4 spent fuel pool—the subject of so much alarm during the crisis—had escaped damage after all (see the appendix for more on this). That finding allowed some in the industry and the NRC to contend that concerns about spent fuel pool risks were unfounded. Their claims did not take into account the fact that U.S. spent fuel pools typically contain several times as much fuel as Unit 4 did, and therefore could experience damage much more quickly in the event of loss of cooling or a catastrophic rupture. The industry argued in turn that FLEX was designed to provide emergency cooling of spent fuel pools. But that would require operator actions at a time when attention would likely be torn by other exigencies, as happened at Fukushima.

Accelerating the transfer of spent fuel to dry casks would enhance safety by passive means. The NRC pledged to look at this issue again, but it was accorded a low priority.

The second and third issues related to emergency planning zones and to distribution of potassium iodide tablets to the public to reduce the threat from radioactive iodine. Fukushima demonstrated that a severe accident could cause radiation exposures of concern to people as far as twenty-five miles from the release site. Indeed, the worst-case projections of both the Japanese and U.S. governments found that even Tokyo, more than one hundred miles away, was similarly threatened.

Nonetheless, the NRC continued to defend the adequacy of a ten-mile planning zone for emergency measures such as evacuation and potassium iodide distribution. The agency argued that evacuation zones could always be expanded if necessary during an emergency—a position hard to reconcile with the likely difficulty of achieving orderly evacuations of people who had not previously known they might need to flee from a nuclear plant accident one day. (Prodded into action by a petition from an activist group, the Nuclear Information and Resource Service, the NRC agreed to look into the emergency planning question, but it is on a very slow track.)

One final issue to consider is the risk of land contamination, something that could be an enormous problem even if all evacuation measures were successful. In the past the NRC has assessed potential accident consequences solely in terms of early fatalities and latent cancer deaths, but Fukushima showed that widespread land contamination, and the economic and social upheaval it creates, must also be counted.[3]

The NRC staff proposed to the commissioners that the NRC address this issue in revising its guidelines for calculating cost-benefit analyses, but the commissioners did not show much interest. It, too, appears to be on a slow track.

As for the fate of the NTTF's Recommendation 1—revising the regulatory framework? That seems to have slipped not only off the front burner but possibly off the stove. In February 2013, the NRC staff failed to meet the commissioner's deadline for a proposal related to the recommendation and asked for more time. As of this writing, Recommendation 1 continues to fade as a priority, with the NRC staff contending that "it is acceptable, from the standpoint of safety, to maintain the existing regulatory processes, policy and framework."

Safety IOUs are worse than worthless. They represent vulnerabilities at operating nuclear plants that the NRC knows to exist but that have not yet been fixed. They are, simply put, disasters waiting to happen.

The NRC's practice of identifying a safety problem and accepting a non-solution continues. The post-Fukushima proposals are just the latest example—albeit the most worrisome.

Severe reactor accidents will continue to happen as long as the nuclear establishment pretends they won't happen. That thinking makes luck one of the defense-in-depth barriers. Until the NRC acknowledges the real possibility of severe accidents, and begins to take corrective actions, the public will be protected only to the extent that luck holds out.

Of course trade-offs are inevitable. It would be ideal if every defense-in-depth barrier were fully and independently protective against known hazards, but realistically that price tag would likely be prohibitive. The nuclear industry is quick to oppose new safety rules on the basis of cost, which is hardly surprising given the concerns of shareholders as well as ratepayers.

On the other hand, one must consider the price tag, in both economic and human terms, of an accident like Fukushima. Ask TEPCO's shareowners—as well as the Japanese public—today what they would have paid to avoid that accident.

So how safe is safe enough? In that critical decision, the public has largely been shut out of the discussion. This is true in the United States, in Japan, and everywhere else nuclear plants are in operation. Nuclear development, expansion, and oversight have largely occurred behind a curtain.

Nuclear technology is extremely complex. Its advocates, in their zeal to promote that technology, have glossed over unknowns and uncertainties, thrown up a screen of arcane terminology, and set safety standards with unquantifiable thresholds such as "adequate protection." In the process, the nuclear industry has come to believe its own story.

Regulators too often have come to believe that there is a firmer technical basis for their decisions than actually exists. Officials, in particular, must grapple with overseeing a technology that few thoroughly understand, especially when things go wrong. Fukushima demonstrated that.

Meanwhile, average citizens have been lulled into believing that nuclear power plants are safe neighbors, needing no attention or concern because the owners are responsible and the regulators are thorough. Yet it is those citizens' health, livelihoods, homes, and property that may be permanently jeopardized by the failure of this flawed system.

The public needs to be fully informed of uncertainties, of risks and benefits, and of the trade-offs involved. Scientists and policy makers must be candid about what they know—and don't know; about what they can honestly promise—and can't promise. And once full disclosures are made, the people must be given the final voice in setting policy. *They* must be the arbiters of what is acceptable and how the government acts to ensure their protection.

What that decision-making process will look like is unclear. One thing is certain: we're nowhere close to it now.

This chapter has focused on two questions inspired by Fukushima: who is to blame for the accident and what can be done to prevent the next one? By now it should be clear that the entire nuclear establishment is responsible, rather than just TEPCO and its regulators. Even if indicted, however, the nuclear establishment likely could not be convicted. For it is sheer insanity to keep doing the same thing over and over hoping that the outcome will be different the next time.

What is needed is a new, commonsense approach to safety, one that realistically weighs risks and counterbalances them with proven, not theoretical, safety requirements. The NRC must protect against severe accidents, not merely pretend they cannot occur.

How can that best be achieved? First, the NRC needs to conduct a comprehensive safety review with the blinders off. That must take place with what former NRC Commissioner Peter Bradford calls regulatory "skepticism." The commission staff—and the five commissioners—should stop worrying so much about maintaining regulatory "discipline" and start worrying more about the regulatory tunnel vision that could cause important risks to be missed or dismissed.

That safety review must come in the form of hands-on, real-time regulatory oversight. Every plant in the United States should undergo the kind of in-depth examination of severe accident vulnerabilities that the NRC contemplated in the 1980s but fell short of implementing.

The first step is adoption of a technically sound analysis method that takes into account the deficiencies in risk assessment that critics have noted over the decades, particularly the failure to fully factor in uncertainty. Issues that are not well understood need to be included in error estimates, not simply ignored. Setting the safety bar at x must carry the associated policy question "What if x plus 1 happens?"

Fortunately, the NRC does not have to start from scratch to do a sound safety analysis. Each nuclear plant that has applied for a twenty-year license renewal from the NRC—around 75 percent of all U.S. plants—has conducted a study called a "severe accident mitigation alternatives" (SAMA) analysis as part of the environmental review required under the National Environmental Policy Act.

A SAFETY FRAMEWORK READY FOR ACTION

A SAMA analysis entails identifying and evaluating hardware and procedure modifications that have the potential to reduce the risk from severe accidents, then determining whether the value of the safety benefits justifies their cost.

Oddly enough, even though the plant owners and the NRC have identified dozens of measures that would pass this cost-benefit test and thus might be prudent investments, none have had to be implemented under the law. That's because the NRC has thrown into the equation its contorted backfit rule. The rule means that for the changes to be required they also must represent a "substantial safety enhancement"—a standard very hard to meet given the low risk estimates generated by the industry's calculations.

Thus, the SAMA process has been merely an academic exercise. But the upgrades identified in the SAMA analyses provide a comprehensive list of changes that could reduce severe accident risk at each plant.

The SAMA changes are a starting point. They should be reevaluated under a new framework, one that better accounts for uncertainties and the limitations of computer models, improves the methodology for calculating costs and benefits, and allows the public to have a say in the answer to the question "How safe is safe enough?" This process would produce a guide for plant upgrades that could fundamentally improve safety of the entire reactor fleet and provide the public with a yardstick by which to measure performance.

Another tool for assessing severe accident risks would be a stress test program—an analysis of how each plant would fare when subjected to a variety of *realistic* natural disasters and other accident initiators. (As for industry's much-touted but untested FLEX "fixes," they could be taken into account, but their limitations and vulnerabilities would be fair game in any analysis.)

Before the testing process begins, another change is essential: the public, not the industry, must first determine what is a passing grade.

These SAMA analyses represent an unvarnished checklist of the changes needed at each nuclear plant in the United States to drive down the risk of an American Fukushima. Using that information as a roadmap for enhanced regulation and operations is not the entire answer, but it is a first step in better understanding the risks of nuclear power and how to control them.

Once the question "How safe is safe enough?" is answered, a second question must be asked and resolved, again with input from the public. That question is: "How much proof is enough"?

In other words, how best to prove that industry and regulators are actually complying with the new rules, both in letter and in spirit? The NRC's regulatory process is among the most convoluted and opaque; just as in Japan, public trust has suffered as a result.

In the end, the NRC must be able to tell the American public, "We've taken every reasonable step to protect you." And it must be the public, not industry or bureaucrats, who define "reasonable."

As Japan was marking the second anniversary of the Fukushima Daiichi accident, the NRC held its annual Regulatory Information Conference, the twenty-fifth, once again attracting a large domestic and international crowd of regulators, industry representatives, and others.

During the two-day session, many presentations were devoted to the lessons learned from Fukushima, including technical discussions about core damage, flooding and seismic risks, and regulatory reforms. But by now it was apparent that little sentiment existed within the NRC for major changes, including those urged by the commission's own Near-Term Task Force to expand the realm of "adequate protection." The NRC was back to business as usual, focused on small holes in the safety net, ignoring the fundamental lesson of Fukushima: This accident should have been no surprise, and without wholesale regulatory and safety changes, another was likely.

One of the final events of the conference was a panel discussion featuring the agency's four regional administrators and two nuclear industry officials, who fielded questions from an audience of other nuclear insiders. The subjects ranged from dealing with the public in the post-Fukushima era to the added workload of inspectors at U.S. reactors. The mood was upbeat, the give-and-take friendly.

But amid the camaraderie, one member of the panel seemed impatient to deliver a message to the audience. When it came to addressing the overarching lessons of Fukushima—for regulators and industry people alike—he brought unique credentials to the task. The speaker was the NRC's own Chuck Casto, who had arrived in Japan in the first chaotic days of the accident and re-mained there for almost a year as an advisor.

Now, as the conference wound to a close, Casto was eager to offer some words of advice. The public does not understand the NRC's underlying safety philosophy of "adequate protection," he cautioned. "They want to see us charging out there making things safer and safer, to be pro-safety. If this degree is safe, a little bit more is *more* safe," he told the audience.

A short time later, his voice filling with emotion, Casto spoke of the brave operators at Fukushima, whom he called "an incredible set of heroes."

"This industry over its fifty-some years has had a lot of heroes," he contin-ued. He spoke about Browns Ferry, Three Mile Island, and Chernobyl. Those and other events, Casto told his audience, make the way forward clear. "We honor and we respect the heroes that we've had in this industry over fifty years—but we don't want any more.

"We have to have processes and procedures and equipment and regulators that don't put people in a position where they have to take heroic action to protect the health and safety of the public," Casto said. "What we really have to work on is no more heroes."

APPENDIX

THE FUKUSHIMA POSTMORTEM:
WHAT HAPPENED?

Accident modelers from TEPCO, the U.S. national laboratories, industry groups, and other organizations gathered in November 2012 at a meeting of the American Nuclear Society (ANS) to present the results of their attempts to simulate the events at Fukushima and reproduce what was known about them.

Like any postmortem, the goal was to glean as many answers as possible about the causes of the events at Fukushima Daiichi. But answers proved troublingly elusive.

One of the first difficulties encountered by the analysts was the lack of good information about the progression of the accidents. The damaged reactors were still black boxes, far too dangerous for humans to enter, much less conduct comprehensive surveys of, and reliable data on their condition was sparse. In some cases, analysts had to fine-tune their models using trial and error, essentially playing guessing games about what exactly had happened within the reactors.

Even so, the computer simulations could not reproduce numerous important aspects of the accidents. And in many cases, different computer codes gave different results. Sometimes the same code gave different results depending on who was using it.

The inability of these state-of-the-art modeling codes to explain even some of the basic elements of the accident revealed their inherent weaknesses—and the hazards of putting too much faith in them.

Sometimes modelers were frustrated by a lack of essential data. For example, when water-level measurements inside the three reactors were available, they were usually wrong. The readings indicated that water levels were stable when they were actually dropping below the bottom of the fuel. This happened because the gauges were not calibrated to account for the extreme temperature and pressure conditions that were occurring. Although the problem should have been obvious at the time, TEPCO didn't question the erroneous data and publicly released it.

The lack of reliable water levels meant that analysts in Japan and elsewhere really did not know then or now how much makeup water was entering the reactor vessels at what times during the accident, a critical piece of information for understanding the effectiveness of emergency water-injection strategies. Different assumptions for "correcting" this unreliable data yielded significantly different results.

At the ANS meeting, researchers from Sandia National Laboratories presented results they obtained using the computer code called MELCOR, designed by Sandia for the NRC to track the progression of severe accidents in boiling water and pressurized water reactors. For Unit 1, the event was close to what's called a "hands-off" station blackout. From the time the tsunami struck, essentially nothing worked. The loss of both AC and battery power disabled the isolation condensers and the high-pressure coolant injection system (HPCI), as well as the instruments for reading pressure and water level. Counting down from the time of the earthquake, the core was exposed after three hours, began to undergo damage after four hours, and by five hours was completely uncovered.

At nine hours, according to this analysis, the molten core slumped into the bottom of the reactor vessel, and by fourteen hours—if not sooner—it had melted completely through. By the time workers had managed to inject emergency water into the vessel at fifteen hours, much of the fuel had already landed on the containment floor and was violently reacting with the concrete.

But even a straightforward "hands-off" blackout turned out to be too complex for MELCOR to fully simulate. For instance, although the code did predict that the containment pressure would rise high enough to force radioactive steam and hydrogen through the drywell seals and into the reactor building, its calculation of the amount of hydrogen that collected at the top of the building "just missed" being large enough to cause an explosion, according to Randy Gauntt, one of the study's authors.

A U.S. industry consultant, David Luxat, presented a simulation using the industry's code, called MAAP5. His simulation also predicted that the conditions would not be right for a hydrogen explosion at the time when one actually occurred. His speculation: extra hydrogen leaked from the vent into the reactor building. Ultimately, the explosion at Unit 1 remained something of a mystery.

Another issue that experts disagreed on was what caused the Unit 1 reactor vessel to depressurize suddenly around six or seven hours into the accident. Sandia argued that it was probably a rupture in one of the steam lines leading from the vessel, or a stuck-open valve; others believed it was a failure of some of the tubes used to insert instruments into the vessel to take readings. But no code was capable of predicting one of these events over another; and no one knew what actually took place anyway. Such confirmation will have to await a time when it is safe to enter the containment and conduct forensic examinations. Even then, it is far from certain that the history of the accident will be fully reconstructed, or all its lessons revealed.

The situation was even murkier when trying to understand the more complex events that led to the meltdowns at Unit 3 and then Unit 2.

For Unit 3, which never fully lost battery power, operators were able to run the reactor core isolation cooling (RCIC) system until it shut down, and then the HPCI system until they deliberately shut it down. Although the analysts generally agreed that core damage occurred sometime before 9:00 a.m. on March 13, there was much disagreement about how extensively the core was damaged and whether it had in fact melted through the reactor vessel. The answers depended on the amount of water that actually got into the vessel from the operations of RCIC, HPCI, fire pumps, and fire engines. Various analysts questioned whether RCIC and HPCI operated well under suboptimal conditions, and whether the pumps ever had sufficient pressure to inject meaningful flows of water into the core. Assuming different amounts of water led to different conclusions. In the final analysis, no one could predict with confidence whether or not there was vessel failure.

The explosion at Unit 3 was another puzzle. It appeared larger than the one at Unit 1, but under the assumptions for water injection rates provided by TEPCO, neither the Sandia simulation nor the industry's found that enough hydrogen was generated to cause any explosion at all.

The analysis of Unit 2 yielded even more mysteries. During the actual accident, Unit 2 was initially a success story compared to its siblings. Even though it lost battery power, the reactor's RCIC system operated for nearly three days. Under conventional modeling assumptions, it would have failed around an hour after the eight-hour batteries ran out. Therefore, the analysts had to force the RCIC system to keep operating under abnormal conditions in their models, even though they didn't understand why. However, they also didn't know how well it had worked. After the RCIC system failed, workers took several hours to start seawater injection; by then, core damage was well under way. As with Unit 3, there was so much uncertainty about the amount of water that reached the core that no simulation could predict with any degree of confidence how much fuel was damaged and whether it melted through the vessel.

Also unexplained was the fact that, although Unit 2's RCIC system was transferring heat from the core to the torus, the measured pressure within the torus was not increasing as much as it should have been. In fact, some of the models had to assume there was a hole in the torus just to make sense of the data. This led to speculation that there had indeed been a hole in the torus from early on in the accident or that the torus room had flooded with water from the tsunami, which helped to cool it down.

If it was a hole, it would have had to have formed well before workers heard the mysterious noise at Unit 2 on the morning of March 15 that made them think an explosion had damaged the torus. That was the only way to explain the pressure data. In any event, TEPCO later concluded that the noise didn't originate at Unit 2 at all, but was an echo of the explosion at Unit 4 that occurred at approximately the same time. The rapid pressure drop that instruments recorded in the Unit 2 torus was attributed by TEPCO to instrument failure. A remote inspection made months after the accident did not reveal any visible damage to the torus.

So Unit 2 was apparently the only unit operating at the time of the earthquake that did not experience a violent hydrogen explosion. One possible reason for this is in the "silver lining" category: the opening created in the Unit 2 reactor building by the Unit 1 explosion prevented the buildup of an explosive concentration of hydrogen gas.

As for the Unit 4 explosion, most evidence indicates that the culprit was not hydrogen released from damaged fuel in the spent fuel pool. In fact, the pool was never as seriously in distress as the NRC, TEPCO, and many people around the world feared.

Inspections of the spent fuel in the pool using remote cameras did not see the kind of damage that would have been apparent if any fuel assemblies had overheated and caught fire. And samples of water from the pool did not reflect the radioactivity concentrations that would accompany damaged fuel. TEPCO was so sure of this finding that it pursued another theory for the explosion: hydrogen had leaked into the Unit 4 building during the venting operations in Unit 3 next door.

Like Units 1 and 2, Units 3 and 4 shared a stack for gas exhaust, and some of the piping from the two units was interconnected. Although closed valves ordinarily would have kept one reactor's exhaust from getting into the other, the valves were not operating properly as a result of the blackout. TEPCO found further evidence for this theory when it measured higher radiation levels on the downstream side of the gas filters at Unit 4 than on the upstream side, indicating that radioactive gas flowed from the outside of the building to the inside.

This didn't mean, however, that the Unit 4 pool was never in danger. TEPCO believed that the pool had lost up to ten feet of water early in the accident as the result of sloshing or some other unknown mechanism, causing the water temperature to shoot up toward the boiling point. Fortuitously, there was a second large pool of water—the reactor well—sitting on top of the reactor vessel. The reactor well is connected to the spent fuel pool through a gate and is filled during refueling to keep fuel rods submerged at all times. When the water in the spent fuel pool dropped, water flowed from the well into the pool, buying time until an effective external water supply was established.

The Unit 2 spent fuel pool may not have been as lucky. It was not one of the pools that commanded much attention during the early days of the accident. Yet when sampling of the pool water was conducted in April 2011, higher than expected levels of radioactivity were detected. Even more startling was a relative absence of iodine-131 in the pool water compared to the amount of cesium-137. If the radioactivity had originated in the reactor cores, more iodine-131 would have been detected. This led officials in the White House Office of Science and Technology Policy to conclude that there was either mechanical or thermal damage to some of the spent fuel in the pool.

Given all the uncertainty, there is little wonder that analysts still do not know exactly how radioactivity was released into the environment, when it was released, and where it came from. There were multiple and sometimes overlapping periods of radioactive releases from the different units.

Most experts agree that large releases on March 14 and 15, coupled with precipitation, were ultimately responsible for the extensive area of contamination stretching northwest from the plant to Iitate village. TEPCO has argued that the venting operations did not contribute significantly to radiation releases and therefore that the reactor toruses were effective in scrubbing fission products from the steam that was vented. The company claims that Unit 2 was the source of the largest release, coming not from the torus through a hole that may or may not exist but from the drywell, which underwent a rapid drop in pressure during the day on March 15. In this view, even the reactor building explosions at Units 1 and 3, as dramatic as they appeared, did not contribute as much to off-site releases, mainly because the buildings could do little to contain radiation even when they were intact.

However, other analysts have looked at the same data and concluded that the venting did cause large releases and that scrubbing in the torus was not effective. This is a crucial technical issue for the U.S. debate over whether filters should be installed on the hardened vents at Mark I and Mark II BWRs. Until forensic investigations narrow down the various possibilities, though, all of these claims remain in the realm of speculation.

Ultimately, based on off-site measurements and meteorological data, it appears that Fukushima Daiichi Units 1 through 3, on average, released to the atmosphere less than 10 percent of the radioactive iodine and cesium that the three cores contained. That would be generally consistent with the results from computer modeling of station blackouts in studies like SOARCA. But there are so many unexplainable features of the accident right now that the similarity in results may be mere coincidence.

What is clear is that, in terms of the amount of radiation released, the Fukushima Daiichi accident was far from a worst-case event. This meant that the direst scenarios that the National Atmospheric Release Advisory Center (NARAC) estimated for Tokyo and parts of the United States, based on much higher radiation releases, never occurred. Fukushima will not challenge Chernobyl's ranking as the world's worst nuclear plant accident in terms of radioactive release, although it will remain classified a level 7 accident by the IAEA.

The difficulties analysts have explaining what happened in 2011 at Fukushima are only compounded when they use the same computer models to predict the future. In other words, when computer models cannot fully explain yesterday's accident, they cannot accurately simulate tomorrow's accident. Yet the nuclear establishment continues to place ever-greater reliance on these codes to develop safety strategies and cost-benefit analyses.

GLOSSARY

AC (alternating current) power: The most common form of electrical power, such as that provided to household appliances plugged into wall outlets. Operating nuclear power plants supply AC power to an electrical grid for use by residential and industrial customers.

Advisory Committee on Reactor Safeguards: An independent committee that advises the U.S. NRC. The committee is made up of experts in a variety of fields and is responsible for reviewing NRC safety studies, license applications, and advising on proposed standards.

B.5.b: Designation for orders issued by the NRC after the September 2001 terrorist attacks, requiring additional safety equipment to be available as backup in the event of a fire or explosion caused by an airplane crashing into a nuclear facility. B.5.b is the title of a section within the NRC's Order for Interim Safeguards and Security Compensatory Measures of February 2002.

beyond-design-basis accidents: Also called "severe" accidents. Accident sequences that are possible but that were not fully considered in the reactor design process because they were judged to be too unlikely (see **design basis**).

boiling water reactor (BWR): A nuclear power reactor design in which water flows upward through the core, where it is heated and allowed to boil in the reactor vessel. The resulting steam then drives turbines, which activate generators to produce AC power.

cold shutdown: A reactor coolant system at atmospheric pressure and at temperature below 200 degrees Fahrenheit following a reactor cooldown.

common-mode failure: Multiple failures of structures, systems, or components as a result of a single phenomenon.

containment building: A steel-reinforced concrete structure that encloses a nuclear reactor. It is intended to minimize the release of radiation in the event of a design-basis accident. In a boiling water reactor, the containment consists of two parts: the drywell and the wetwell.

control rod: A rod that is used to control power output by controlling the nuclear chain reaction rate inside a reactor. This rod contains material that easily absorbs neutrons and can suppress nuclear fission when inserted between the fuel assemblies. In an emergency, all control rods fully insert into the reactor core to stop the nuclear chain reaction.

core: See **reactor core**.

cost-benefit analysis: A systematic economic evaluation of the positive effects (benefits) and negative effects (non-benefits, including monetary costs) of undertaking an action.

criticality: The normal operating condition of a reactor where the nuclear chain reaction is sustained. A reactor achieves criticality (and is said to be critical) when each fission event releases a sufficient number of neutrons to sustain an ongoing series of reactions.

DC (direct current) power: The electrical current produced by batteries and from inverters that transform AC power to DC power.

decay heat: When an atomic nucleus fissions, two smaller nuclei are commonly formed. Many of these fission by-products are unstable and release radiation seeking to become stable. These radioactive emissions generate thermal energy called decay heat. Even after a reactor core is shut down and its nuclear chain reaction stopped, it continues to generate substantial amounts of decay heat that can cause fuel damage from overheating if not removed at a sufficiently high rate.

defense-in-depth: Multiple independent and redundant layers of defense to compensate for potential human and mechanical failures so that no single layer is exclusively relied upon. For example, the array of emergency pumps installed to cool the reactor core during an accident, the containment building that minimizes radiation escaping from a damaged reactor core, and the emergency plans that evacuate people in event of a radiation release are each defense-in-depth layers.

design basis: The range of conditions and events taken explicitly into account in the design of a facility, according to established criteria, so that the facility can withstand them without exceeding authorized limits by the planned operation of safety systems.

dose: A measure of the energy deposited by radiation in a target.

dosimeter: A small portable instrument used to measure and record the total accumulated dose of ionizing radiation.

dry cask storage: A passive means of storing reactor fuel that provides radiation shielding. After several years in a spent fuel pool, fuel can be transferred to a cask, typically made of steel, concrete and other materials. Once in the casks, the fuel is cooled by natural airflow.

drywell: In a boiling water reactor with a Mark I containment, the drywell houses the reactor vessel. It resembles an inverted incandescent lightbulb and is made of a steel shell surrounded by steel-reinforced concrete.

emergency core cooling systems (ECCS): Reactor system components (pumps, valves, heat exchangers, tanks, and piping) specifically designed to remove decay heat from the reactor core in the event of a failure that drains or leaks the normal cooling water.

emergency diesel generators: Equipment permanently installed at nuclear plant sites that burns diesel fuel oil to generate AC power to supply emergency core cooling systems and other emergency equipment when off-site power is unavailable.

emergency planning zone: An area of approximately ten-mile radius surrounding a nuclear power plant where the principal exposure sources would be whole body external exposure to gamma radiation from the plume and deposited material, and inhalation exposure from the passing radioactive plume.

external event: Potentially damaging events originating from outside a nuclear plant site or outside of safety-related buildings. Typical examples of external events for nuclear facilities include earthquakes, tornadoes, tsunamis, and aircraft crashes. The NRC also considers plant fires to be "external events" even if they originate within plant buildings.

feed and bleed: A process in which makeup water is added to a reactor vessel when the closed-loop cooling system is malfunctioning. The makeup water absorbs heat given off by the nuclear fuel and is allowed to boil, or "bleed," away.

FLEX: Short for "diverse and flexible mitigation capability," the program was devised by the U.S. nuclear industry in the months following the Fukushima Daiichi accident. It comprises a variety of portable equipment that can be rapidly installed in or deployed to a nuclear facility in the event of an accident or natural disaster to provide backup electrical and cooling systems.

fuel assembly: A set of fuel rods loaded into and subsequently removed from a reactor core as a single unit.

fuel pellet: In light-water reactors, a thimble-sized ceramic cylinder (approximately ⅜-inch in diameter and ⅝-inch in length), consisting of uranium (typically uranium oxide, UO_2), which has been enriched to increase the concentration of uranium-235 (U-235). Reactor cores may contain up to 10 million pellets, stacked in the fuel rods and arranged in fuel assemblies.

fuel rod: A hollow tube more than twelve feet long made from a metal alloy containing zirconium that is filled with fuel pellets and included in a fuel assembly.

hardened vent: A metal pipe designed to withstand the higher pressure that may occur inside containment during an accident that can be used to discharge (vent) the containment atmosphere to an elevated point. The normal ventilation system for lower pressures uses sheet-metal ducts similar to those used in homes and offices.

heat sink (also **ultimate heat sink**): The nearby river, ocean, or lake used to cool the nuclear plant. During normal operation, one unit of electricity is generated and two units of waste heat are rejected to the heat sink for every three units of thermal power produced by the reactor core. During accidents after a successful scram, a smaller amount of water from the ultimate heat sink is required to cool the reactor core and emergency equipment.

high-pressure coolant injection (HPCI) system: Part of the emergency core cooling systems. HPCI is designed to inject substantial quantities of water into the reactor while it is at high pressure. Like the reactor core isolation cooling system (RCIC), it can run without AC power as long as DC power is available. HPCI uses steam produced by the reactor core's decay heat to spin a turbine connected to a pump.

hypocenter: The point of the earth's crust where a rupture initiates, creating an earthquake.

International Nuclear and Radiological Event Scale (INES): A scale developed by the International Atomic Energy Agency to communicate in a consistent way the safety significance of nuclear and radiological events. Events are classified on a scale with seven levels, ranging from level 1 (an "anomaly") with little danger to the general population, to a level 7 (a "major accident") with a large release of radioactive materials and widespread health and environmental effects. The Chernobyl and Fukushima nuclear accidents were both level 7 accidents.

MELCOR: A computer code developed for the NRC by Sandia National Laboratories. It models the progression of severe accidents in boiling water and pressurized water reactors.

meltdown: Large-scale melting of nuclear fuel rods in the reactor core.

millirem: One thousandth of a rem (0.001 rem). A measure of radiation dose. (See **rem.**)

millisievert: One thousandth of a sievert (0.001 sievert). A measure of radiation dose (see **sievert**).

Nuclear Regulatory Commission, U.S. (NRC): An independent agency created by the Energy Reorganization Act of 1974, replacing part of the former Atomic Energy Commission (AEC). The NRC is made up of five commissioners appointed by the president, who serve staggered, five-year terms. One member serves as chairman and acts as principal executive officer. The commission formulates policies, develops regulations governing nuclear reactor and nuclear material safety, issues licenses and is responsible for overseeing reactor and nuclear material safety.

off-site power: Power supplied to a power station via transmission lines from the electrical power grid.

operating basis earthquake: The earthquake that, considering the regional and local geology and seismology, and specific characteristic of local subsurface material, could reasonably be expected to affect a nuclear plant during its operating life.

potassium iodide: A compound that provides protection to the thyroid from exposure to radioactive iodine-131, which is one of the fission products that can be released in a meltdown. Potassium iodide would not give any protection against any other radioactive isotope that may be released in a meltdown at a nuclear power plant.

primary containment: See **containment building**.

pressurized water reactor (PWR): A nuclear power reactor design that utilizes two separate coolant loops, unlike a boiling water reactor. In the primary loop, water is circulated through the reactor core and heated to a very high temperature by fission, but kept under high pressure to prevent it from boiling. A PWR essentially operates like a pressure cooker, where a lid is tightly placed over a pot of heated water, causing the pressure inside to increase as the temperature increases (because the steam cannot escape) but keeping the water from boiling at the usual 212°F (100°C). The heated water is piped through tubes called steam generators, which transfer heat energy to water in a secondary loop, where it is allowed to boil. The resulting steam is used to drive turbine generators to produce electrical power. About two-thirds of the operating nuclear power plants in the United States are PWRs.

probabilistic risk assessment (PRA): A systematic method for assessing three questions that the NRC uses to define "risk." These questions consider (1) what can go wrong, (2) how likely it is, and (3) what its consequences might be. The NRC uses PRA to determine a numeric estimate of risk to provide insights into the strengths and weaknesses of the design and operation of a nuclear power plant.

Protective Action Guides (PAGs): Suggested precautions that state and local authorities can take during an emergency to keep people from receiving an amount of radiation judged excessively dangerous to their health. The U.S. Environmental Protection Agency developed the PAG Manual to provide guidance on actions to protect the public, such as having people evacuate an area or stay indoors.

RASCAL (Radiological Assessment System for Consequence Analysis): A computer code used by the U.S. NRC to make dose projections for atmospheric releases out to fifty miles during radiological emergencies.

reactor building: Also called the secondary containment, the structure of a boiling water reactor that completely surrounds the containment building. The reactor building houses the emergency core cooling systems and the spent fuel pool (at most boiling water reactors).

reactor core: The central portion of a nuclear reactor that contains the fuel assemblies, moderator, control rods, and support structures. The reactor core is where fission occurs.

reactor coolant system: The reactor vessel (the metal pot housing the reactor core), the piping and components attached to it, and the water contained inside that function to remove heat produced by the reactor core during normal operation and accidents.

reactor core isolation cooling (RCIC) system: A system designed to supply makeup water to a shutdown reactor to compensate for boil-off caused by decay heat when the normal makeup system has failed or is otherwise unavailable. Like HPCI, RCIC uses steam produced by the reactor core's decay heat to spin a turbine connected to a pump and can run in a stable fashion as long as DC power is available to open its valves and control its turbine speed.

rem (roentgen equivalent man): A unit primarily used in the United States to measure the dose equivalent (or effective dose), which combines the amount of energy (from any type of ionizing radiation that is deposited in human tissue), along with the medical effects of the given type of radiation. The related international system unit is the sievert (Sv), where one hundred rem is equivalent to one Sv.

relief valve (also safety relief valve): Valves installed on reactor coolant system pipes for protection against damage caused by high internal pressure. The valves have springs that keep them closed normally. If pressure inside the pipe rises too high or the operators apply compressed air pressure, a relief valve will open to reduce pressure by allowing reactor coolant system fluid to flow into the wetwell of a boiling water reactor.

safe shutdown earthquake (SSE): The maximum earthquake potential for which certain structures, systems, and components, important to safety, are designed to sustain and remain functional.

Severe Accident Management Guidelines (SAMGs): A set of voluntary nuclear industry guidelines of the 1990s to be used by plant operators in the event of a beyond design-basis accident in which core damage had already occurred or was imminent.

shutdown: A decrease in the rate of fission (and heat/energy production) to reach a subcritical state in a reactor (usually by the insertion of control rods into the core).

sievert (Sv): An international unit of measurement for radiation dosage (1 Sv = 100 rem). According to the World Health Organization, the average person is exposed to about three millisieverts a year of radiation, from naturally occurring, medical, and other sources. Exposure to about one thousand millisieverts (one sievert) of radiation in a short period of time can cause acute radiation sickness.

State-of-the-Art Reactor Consequence Analyses (SOARCA): A study begun in 2007 by the NRC to develop estimates of off-site radiological health consequences for severe reactor accidents.

source term: The amount and isotopic composition of material released (or postulated to be released) from a facility. Used in modeling releases of radionuclides to the environment, particularly in the context of accidents at nuclear installations or releases from radioactive waste in repositories.

SPEEDI: System for Prediction of Environmental Emergency Dose Information. Japan's national monitoring network to measure real-time dose assessments during radiological emergencies. Developed by the Japanese after the Three Mile Island accident, it began operations in 1986.

spent fuel pool: A pool of reinforced concrete typically about forty feet deep in which fuel assemblies are stored beneath at least twenty feet of water. The water, which circulates to provide cooling, also acts to shield radiation. These pools can be located above the reactor, as is the case with boiling water reactors, or in an adjacent ground-level area, as is the case with pressurized water reactors.

spent nuclear fuel: Nuclear reactor fuel that has been used to the extent that it can no longer effectively sustain a chain reaction.

station blackout: A complete loss of alternating current electric power to the station.

suppression chamber (and suppression pool): See **wetwell**.

torus: See **wetwell**.

turbine building: The building housing the turbine/generator and auxiliary equipment.

vent valves (also containment vent valves): A series of valves that factored significantly in the Fukushima accident. The containment vent valves remain closed during normal operation and during design basis accidents. During beyond design-basis accidents, the containment vent valves may be opened to allow gases to be discharged to control hydrogen buildup, remove heat, or lower pressure.

wetwell: In a boiling water reactor, the wetwell sits below the drywell and is connected to it through a series of pipes. The wetwell is sometimes called a torus, suppression chamber or suppression pool. It contains a large volume of water that acts as an "energy sponge" to absorb energy released from the reactor coolant system during an accident.

KEY INDIVIDUALS

In Japan

Shinzo Abe, prime minister (December 2012–)

Yukiya Amano, director, International Atomic Energy Agency

Yukio Edano, chief cabinet secretary (January 2011–September 2011); minister of economy, trade, and industry (September 2011–December 2012)

Toru Hashimoto, mayor, Osaka

Yotaro Hatamura, chairman, Investigation Committee on the Accident at the Fukushima Nuclear Power Stations of Tokyo Electric Power Company

Naomi Hirose, president, TEPCO (June 2012–)

Goshi Hosono, minister of the environment (September 2011–December 2012)

Katsuhiko Ishibashi, seismologist and nuclear safety proponent

Banri Kaieda, minister of economy, trade, and industry (January 2011–August 2011)

Naoto Kan, prime minister (June 2010–August 2011)

Tsunehisa Katsumata, chairman, TEPCO (June 2008–May 2012)

Toshiso Kosako, radiation safety expert, advisor to Prime Minister Kan

Haruki Madarame, chairman, Nuclear Safety Commission

Kiyoo Mogi, seismologist and nuclear safety proponent

Koichiro Nakamura, deputy director general, Nuclear and Industrial Safety Agency

Toshio Nishizawa, president, TEPCO (June 2011–May 2012)

Yoshihiko Noda, prime minister (September 2011–December 2012)

Masataka Shimizu, president, TEPCO (June 2008–June 2011)

Kazuhiko Shimokobe, chairman, TEPCO (April 2012–)

Aileen Mioko Smith, executive director, Green Action

Ichiro Takekuro, TEPCO liaison to Japanese prime minister Kan

Masao Yoshida, site superintendant, Fukushima Daiichi Nuclear Power Plant

Atsufumi Yoshizawa, Fukushima Daiichi employee

In America

Steven Aoki, deputy undersecretary, U.S. Department of Energy

George Apostolakis, commissioner, U.S. Nuclear Regulatory Commission

James K. Asselstine, commissioner, U.S. NRC (1982–1987)

Bill Borchardt, executive director for operations, U.S. NRC

Eliot Brenner, Office of Public Affairs, U.S. NRC

Scott Burnell, Office of Public Affairs, U.S. NRC

Charles Casto, deputy regional administrator, U.S. NRC; headed the NRC's team in Japan

Bill Dedman, reporter for NBC News and msnbc.com

Kirkland Donald, director, Office of Naval Reactors, U.S. DOE

Daniel Dorman, deputy director, Office of Nuclear Material Safety and Safeguards, U.S. NRC (later a member of the NRC's Near-Term Task Force)

Jack Grobe, deputy director, Office of Nuclear Reactor Regulation, U.S. NRC (later a member of the NRC's Near-Term Task Force)

Gregory Jaczko, chairman, U.S. NRC (2009–2012); NRC commissioner (2005–2009)

Annie Kammerer, seismologist and earthquake engineer, U.S. NRC

Allison Macfarlane, chairman, U.S. NRC (2012–)

William D. Magwood, commissioner, U.S. NRC

Dr. Charles "Charlie" Miller, director, Office of Federal and State Materials and Environmental Management Programs, U.S. NRC (later the chair of the NRC's Near-Term Task Force)

John Monninger, deputy chief of staff for the chairman, U.S. NRC, and member of the NRC's team in Japan

Michael "Mike" Mullen, chairman, Joint Chiefs of Staff

William C. Ostendorff, commissioner, U.S. NRC

Tony Pietrangelo, chief nuclear officer, Nuclear Energy Institute

John V. Roos, U.S. ambassador to Japan

Brian Sheron, director, Office of Nuclear Regulatory Research, U.S. NRC

Kristine L. Svinicki, commissioner, U.S. NRC

Jim Trapp, branch chief, U.S. NRC, and member of the NRC's team in Japan

Stephen Trautman, deputy director, Office of Naval Reactors, U.S. DOE

Anthony Ulses, chief, Reactor Systems Branch, U.S. NRC, and member of the NRC's team in Japan

Martin Virgilio, deputy executive director, Reactor and Preparedness Programs, U.S. NRC

Mike Weber, deputy executive director, U.S. NRC

Jim Wiggins, director, Office of Nuclear Security and Incident Response, U.S. NRC

U.S. BOILING WATER REACTORS WITH "MARK I" AND "MARK II" CONTAINMENTS

REACTORS	CONTAINMENTS	LOCATION
Browns Ferry 1	BWR-MARK I	Alabama
Browns Ferry 2	BWR-MARK I	Alabama
Browns Ferry 3	BWR-MARK I	Alabama
Brunswick 1	BWR-MARK I	North Carolina
Brunswick 2	BWR-MARK I	North Carolina
Cooper	BWR-MARK I	Nebraska
Dresden 2	BWR-MARK I	Illinois
Dresden 3	BWR-MARK I	Illinois
Duane Arnold	BWR-MARK I	Iowa
Fermi 2	BWR-MARK I	Michigan
FitzPatrick	BWR-MARK I	New York
Hatch 1	BWR-MARK I	Georgia
Hatch 2	BWR-MARK I	Georgia
Hope Creek 1	BWR-MARK I	New Jersey
Monticello	BWR-MARK I	Minnesota
Nine Mile Point 1	BWR-MARK I	New York
Oyster Creek	BWR-MARK I	New Jersey
Peach Bottom 2	BWR-MARK I	Pennsylvania
Peach Bottom 3	BWR-MARK I	Pennsylvania
Pilgrim 1	BWR-MARK I	Massachusetts
Quad Cities 1	BWR-MARK I	Illinois
Quad Cities 2	BWR-MARK I	Illinois
Vermont Yankee	BWR-MARK I	Vermont
Columbia Generation Station	BWR-MARK II	Washington
La Salle 1	BWR-MARK II	Illinois
La Salle 2	BWR-MARK II	Illinois
Limerick 1	BWR-MARK II	Pennsylvania
Limerick 2	BWR-MARK II	Pennsylvania
Nine Mile Point 2	BWR-MARK II	New York
Susquehanna 1	BWR-MARK II	Pennsylvania
Susquehanna 2	BWR-MARK II	Pennsylvania

Source: U.S. Nuclear Regulatory Commission.

NOTES AND REFERENCES

The following are key sources and explanatory notes for *Fukushima: The Story of a Nuclear Disaster*. Full references are available at Fukushimastory.com, including links to documents, reports, illustrations, and other materials that played a role in the preparation of this book.

The sources for this book include contemporaneous news accounts published in the United States, Japan, and elsewhere. Extensive use was made of information gathered by several investigations into the Fukushima Daiichi accident that were conducted by official Japanese panels, international industry associations, scientific groups, and the plant's owner, the Tokyo Electric Power Company, as well as analyses by nuclear safety organizations, notably the Union of Concerned Scientists.

Playing a major role in this book are transcripts maintained by the Nuclear Regulatory Commission of conversations among NRC experts in the United States and in Japan. These transcripts, plus thousands of pages of NRC documents and e-mails related to the commission's response during and after the accident, provide an inside view of the challenges and uncertainties facing those charged with protecting public health and safety.

The book also draws on the long legislative and regulatory histories of the Nuclear Regulatory Commission and its predecessor, the Atomic Energy Commission. Public transcripts of meetings, written opinions, congressional hearings, and scientific reports were also incorporated.

1. March 11, 2011: "A Situation That We Had Never Imagined"

News accounts of the accident provided descriptions of the early hours of the accident. The Earthquake Engineering Research Institute's reports on the Tohoku earthquake (available at www.eeri.org/2011/03/tohoku-japan/) offered valuable information, as did two articles published in the Bulletin of the Atomic Scientists, *"Fukushima: The Myth of Safety, the Reality of Geoscience" (September–October 2011) and "Fukushima in Review: A Complex Disaster" (March 8, 2012). Additional material came from the Investigation Committee on the Accident at Fukushima Nuclear Power Stations of Tokyo Electric Power Company (www.cas.go.jp/jp/seisaku/icanps/eng/), hereinafter the "Hatamura report," and the Institute of Nuclear Power Operations, "Special Report on the Nuclear Accident at the Fukushima Daiichi Nuclear Power Station," November 2011 (available at www .nei.org/corporatesite/media/filefolder/11_005_Special_Report_on_Fukushima_Daiichi _MASTER_11_08_11_1.pdf).*

1. The Richter scale for earthquake magnitude has been replaced among scientists by the moment magnitude scale, which is now the most commonly cited measure for medium to large earthquakes. Although it is not identical to the Richter scale,

the moment magnitude scale provides comparable numbers, and they are the ones used in this book. The moment magnitude scale is a logarithmic scale in which each unit of increase in magnitude represents a thirty-two-fold increase in the total energy released. Scientists also use the Modified Mercalli Intensity Scale, which measures the effects of an earthquake rather than its magnitude. This scale has twelve levels of intensity ranging from imperceptible shaking (I) to catastrophic destruction (XII). The Japan Meteorological Agency (JMA) uses yet another scale ranging from I to VII, the highest category. On March 11, 2011, much of the east coast of Japan received Intensity VII shaking, about the same as Modified Mercalli Intensity X.

2. Each control rod (CR) is equipped with position indicators. When all the control rod indicators reflect the fully inserted position, this alarm saves the operators the trouble of having to check each control rod's status.

3. Alternating current (AC) electricity powers home appliances plugged into electrical outlets. Direct current (DC) electricity powers portable electronic devices with batteries. Many of the components of emergency systems at nuclear power plants, like motor-driven pumps and valves, need AC electricity to run. DC power from batteries can be used for the instrumentation and control systems that operate systems like the isolation condensers and RCIC.

4. The word *tsunami* was introduced to English speakers by two journalists working in Japan at the time of the 1896 disaster. Eliza Ruhamah Scidmore, an American, wrote an account of the June 15 earthquake and tsunami for the September 1896 issue of *National Geographic* magazine: "There were old traditions of such earthquake waves on this coast," she wrote. Scidmore would go on to become the first female board member of the National Geographic Society and play a role in planting cherry trees along the Potomac River. The second journalist to describe the disaster was Lafcadio Hearn. Writing in the December 1896 *Atlantic Monthly*, Hearn reported: "From time immemorial the shores of Japan have been swept, at irregular intervals of centuries, by enormous tidal waves.... These awful sudden risings of the sea are called by the Japanese *tsunami*."

5. The U.S. Geological Survey has officially designated it the Great Tohoku Earthquake.

6. The risk of station blackout had been raised within the Nuclear Regulatory Commission in the 1970s. The nuclear industry resisted this concern, asserting that the likelihood of the electrical grid and the on-site emergency diesel generators all failing simultaneously was low. That argument prevailed until March 1990, when the Unit 1 reactor at the Vogtle plant in Georgia actually experienced a blackout. Fortunately, the plant was shut down when the blackout occurred, giving workers sufficient time to repair the emergency diesel generator and avert disaster.

7. Fukushima Daiichi is loosely translated as "Fukushima One" and Fukushima Daini as "Fukushima Two." The designation is based on their construction dates.

8. Japan has two incompatible electricity grids. The grid supplying the northeastern half of the country, including an area served by TEPCO, operates at fifty hertz. Power to the western portion of the island is sixty hertz. The two grids arose from the original sources of generation equipment: Germany (fifty hertz) and the United States (sixty hertz). The inability to share large amounts of electricity between the grids

compounded energy shortages during the Fukushima accident and can do so at other times when demand peaks in one area but not the other.

2. March 12, 2011: "This May Get Really Ugly . . ."

Transcripts of conversations recorded at the NRC's emergency operations center as well as e-mails and other NRC documents were used in this chapter. News accounts, scientific journal articles, and information from the U.S. Geological Survey were utilized. Details of the accident at the Tokai nuclear facility were published in "Accident Prone: The Trouble at Tokai-Mura," Bulletin of the Atomic Scientists, March–April 2000. Information also comes from the Hatamura report.

1. For disclosure, one of the authors (Lochbaum) has traveled through the "revolving door": having worked for more than a decade in the U.S. nuclear power industry, he moved to the Union of Concerned Scientists (UCS) for more than a decade, went to the NRC for more than a year, then returned through the door to the UCS.

2. As a result of the Kashiwazaki-Kariwa fire, TEPCO required that fire engines be stationed at its nuclear plants, which proved beneficial at Fukushima Daiichi in the days after March 11.

3. March 12 Through 14, 2011: "What the Hell Is Going On?"

The Hatamura investigation, NRC transcripts, and the Institute for Nuclear Power Operations' Special Report on the Nuclear Accident at the Fukushima Daiichi Nuclear Power Station, INPO 11-005 (Atlanta: Institute of Nuclear Power Operations, 2011), www.nei.org/corporatesite/media/filefolder/11_005_Special_Report_on_Fukushima _Daiichi_MASTER_11_08_11_1.pdf, were used in this chapter.

1. If even that limited information had been heeded, investigators later concluded, evacuation routes for several communities northwest of the reactors could have been altered to avoid exposing the public to the radiation plumes they thought they were fleeing.

2. Water acts as a "moderator" in reactors like those at Fukushima, slowing neutrons so that they more readily split atoms. Fuel, water, and control rods are carefully arranged to ensure that the nuclear chain reaction can be controlled. But injecting water into the potentially molten core carried the risk of increasing the fission rate and possibly restarting a chain reaction. For this reason, water reserves maintained at reactors are mixed with boric acid, a neutron absorber, before being injected into the core. The boric acid can help prevent a chain reaction.

3. Those standards are set by the Environmental Protection Agency and known as Protective Action Guides, or PAGs.

4. A hardened vent is a separate vent pipe capable of withstanding heavier loads during an accident, such as a station blackout. It is a safety mechanism for relieving pressure in the primary containment by releasing steam and radioactivity at an elevated point outside the reactor building.

5. In the event of a nuclear plant disaster involving radiation releases in the United States, the states, not federal agencies, decide on public protection measures

like evacuations and sheltering. The NRC's role is to provide information to state authorities to inform their ultimate decisions.

6. In pressurized water reactors, the spent fuel pools are located in buildings adjacent to and outside of the containment. This separation makes it less likely that reactor issues will affect spent fuel pools and vice versa. The spent fuel pools are at or below ground level; this placement makes it easier to add water, because smaller pumps are needed to move water across the ground than to push it up five floors. But even the PWR spent fuel pools typically have space below them where water could quickly drain if the pool liner were breached.

7. During normal operation, steam produced inside the reactor vessel is transported through four large pipes. Each pipe is equipped with three or four relief valves, which automatically open when pressure rises or when pressure risks are too high. The relief valves can also be manually opened from the control room—when power is available—to allow the operators to reduce pressure in the steam lines and in the reactor vessel itself. When a relief valve is open, steam flows via a metal pipe into the torus, where it is discharged beneath the surface of the water. Thus, the heat energy from the steam remains inside the containment. The containment vent valves, in contrast, transfer energy to the atmosphere outside.

8. The U.S. NRC does not limit the amount of radiation workers may receive in an emergency. The U.S. Environmental Protection Agency has set voluntary guidelines of up to twenty-five rem (250 millisieverts) for life-threatening emergencies.

4. March 15 Through 18, 2011: "It's Going to Get Worse . . ."

NRC transcripts, congressional hearing testimony, and the Hatamura report provided information for this chapter, as well as testimony delivered to subcommittees of the House Committee on Energy and Commerce, "The Fiscal Year 2012: Department of Energy and Nuclear Regulatory Commission Budgets," Joint Hearing Before the Subcommittee on Energy and Power and the Subcommittee on Environment and the Economy, 112th Cong., March 16, 2011, www.gpo.gov/fdsys/pkg/CHRG-112hhrg68480/pdf/CHRG-112 hhrg68480.pdf, and to the Senate Environment and Public Works Committee, "Full Committee Briefing on Nuclear Plant Crisis in Japan and Implications for the United States," March 16, 2011, www.epw.senate.gov/public/index.cfm?FuseAction=Hearings .Hearing&Hearing_ID=bb6c78e6-802a-23ad-4c7b-9aa7a3bb0c31.

1. Masao Yoshida died of esophageal cancer on July 9, 2013, at age fifty-eight. In November 2011, after being diagnosed with the disease, he took a leave of absence from TEPCO. Medical experts said his cancer was unrelated to radiation exposure from the accident at Fukushima Daiichi.

2. Casto had just three hours to pack before boarding the plane bound for Narita Airport. Flight personnel, noticing Casto's shirt with its NRC logo, soon relocated him to a seat in the nearly empty first-class section and began peppering him with questions about the accident in Japan. Uppermost in their minds: was it safe to even fly into the country? Upon arriving in Tokyo, Casto soon discovered that his choice of clothing—what he calls "emergency response casual," polo shirts and khakis—was not

suitable for his formal meetings with the Japanese. He called his wife and asked her to FedEx every suit in his closet.

3. The NRC's Office of Research had neglected to provide Casto with more recent technical analyses from the agency's State-of-the-Art Reactor Consequences Analyses program. These analyzed accident scenarios very similar to those occurring at Fukushima in even more detail.

4. B.5.b is the title of a section within the NRC's *Order for Interim Safeguards and Security Compensatory Measures* of February 2002.

5. The role played by Japan's Self-Defense Forces (SDF) was as symbolic as it was material. About one hundred thousand SDF troops had been called up by the government on March 11 to respond to the natural disaster. As conditions at Fukushima Daiichi deteriorated, the SDF mission was broadened to help with the reactor crisis. "The SDF operation [at Fukushima Daiichi] encouraged others . . . to say that this problem was not just TEPCO's accident, it was Japan's," Prime Minister Kan later explained. "Everyone started to feel strongly that this was about whether Japan would survive."

5. Interlude—Searching for Answers: "People . . . Are Reaching the Limit of Anxiety and Anger"

This chapter drew on documents from the NRC; National Diet of Japan, The Official Report of the Fukushima Nuclear Accident Independent Investigation Commission (Tokyo: National Diet of Japan, 2012), warp.da.ndl.go.jp/info:ndljp/pid/3856371/naiic .go.jp/wp-content/uploads/2012/09/NAIIC_report_lo_res10.pdf; the Hatamura report; and the Tokyo Electric Power Company. In addition, journalist Bill Dedman provided background information and correspondence.

1. Like other Japanese leaders during the first weeks of the crisis, Edano routinely appeared before the cameras in sky-blue coveralls. The two-piece uniform is worn by officials during emergencies to demonstrate a sense of unity with labor crews, reported the *Wall Street Journal.* Similar two-piece company uniforms were worn by TEPCO executives throughout the accident.

2. More than half the people ordered to leave twelve municipalities around the plant remained unaware that their evacuation was the result of problems at Fukushima Daiichi until the morning of March 13, nearly two days after the accident, according to a later survey.

6. March 19 Through 20, 2011: "Give Me the Worst Case"

NRC transcripts and documents were a primary source in this chapter. Also cited is a report titled "Overview of the Department of Energy's Radiological Dose Assessment of the Fukushima Daiichi Nuclear Power Station Releases," prepared by Terry Kraus and Brian Hunt of Sandia National Laboratories, SAND Report 2012-1226C, astarnmjss .nmcourts.gov/speakernotes/Japan_CM_Overview_Brief.pdf.

This chapter also utilized information obtained from a 2012 Union of Concerned Scientists Freedom of Information Act request to the White House Office of Science and Technology Policy.

1. The accuracy of water-level readings was a chronic problem at Fukushima Daiichi. Water level indicators for the reactor cores themselves were erroneous, in addition to the uncertainty about water covering the spent fuel.

2. Once the use of freshwater could be restored, the salt deposits building up inside the reactor would begin to dissolve, allowing cooling water to circulate more effectively.

3. MELCOR is a computer code developed for the NRC by Sandia National Laboratories. It models the progression of severe accidents and can estimate source terms under a variety of accident scenarios.

7. Another March, Another Nation, Another Meltdown

Report of the President's Commission on the Accident at Three Mile Island: The Need for Change: The Legacy of TMI *(Washington, DC: U.S. Government Printing Office, 1979), available at www.threemileisland.org/downloads/188.pdf, provided historical materials in this chapter. Additional details can be found at the NRC's "Backgrounder on the Three Mile Island Accident," www.nrc.gov/reading-rm/doc-collections/fact-sheets/3mile-isle.html.*

1. According to a survey by the Kemeny Commission, the Union of Concerned Scientists led the list of organizations the media relied on during the accident.

8. March 21 Through December 2011: "The Safety Measures . . . Are Inadequate"

Interview with Charles Casto of the NRC, findings of the Hatamura investigation, and NRC documents provided background for this chapter, as well as the article "The Radiological and Psychological Consequences of the Fukushima Daiichi Accident," Bulletin of the Atomic Scientists, *September–October 2011.*

1. Japan's demonstration breeder reactor, Monju, generated power for only a few months before a sodium fire disabled it in December 1995. After plant overseers finally obtained permission to restart it in 2010, another accident shut it down until at least 2014.

9. Unreasonable Assurances

This chapter uses information from numerous articles from the trade newsletters Nucleonics Week *and* Inside N.R.C. *dating to the early 1980s. It draws from numerous public documents from the NRC, General Accounting Office (now the Government Accountability Office), Congressional Research Service, the Federal Register, and TEPCO. It also draws upon two outstanding books:* Citizen Scientist *by Frank von Hippel (New York: Simon & Schuster, 1991), 19–21, and* Tritium on Ice *by Kenneth D. Bergeron (Cambridge, MA: MIT Press, 2001), 57–58.*

1. When the NRC staff is considering "especially important or controversial rules," the agency publishes what it calls an Advance Notice of Proposed Rulemaking, laying out the issue and soliciting comments from the public in advance of developing

a draft rule. Ordinarily public comment would not be sought until after publication of a draft rule.

2. A *backfit* is a modification of or addition to systems, structures, components, or the design of an existing nuclear plant resulting from changes in the NRC's rules or reinterpretations of old ones.

3. Also, by not having to submit a license amendment, owners avoided giving the public an opportunity to intervene and challenge its adequacy. Any "voluntary" upgrades would be conducted outside public scrutiny.

4. When the NRC inspected U.S. nuclear plants shortly after Fukushima, it found that while all plants had SAMGs, 23 percent did not train workers on the tasks assigned to them in the SAMGs, and 40 percent never conducted exercises using the SAMGs to see whether workers could successfully follow the guidelines. In addition, fewer than half the plants updated their SAMGs to reflect modifications made to the plants.

5. Two examples illustrate how the NRC's risk sword was razor sharp on one side and Nerf-like on the other. In the 1990s, two decades of experience found little damage in pipes and the welds holding them together, indicating that the inspection requirements were unnecessarily strict. As a result, the NRC significantly reduced the frequency of pipe and weld inspections. On the other hand, when evidence emerged that ice condensers and the Mark III were vulnerable to containment failure in station blackouts, the NRC refused to strengthen its rules to require backup power for the igniter systems needed to burn off hydrogen.

6. According to a March 2011 e-mail from NRC staff member Martin Stutzke concerning the Indian Point plant, "the licensee told us they does [*sic*] not maintain their seismic PRAs . . . there's little regulatory motivation for them to do so. It's somewhat understandable that the licensee would focus on the internal event CDF [core damage frequency] . . . and tend to ignore the contributions from earthquakes and fires."

7. Through the process of "radiolysis," oxygen would start to build up as radiation broke water apart into oxygen and hydrogen molecules.

10. "This Is a Closed Meeting. Right?"

This chapter is indebted to several contemporaneous stories about the CRAC2 report by Milton Benjamin in the Washington Post *and Robert Sangeorge of United Press International, as well as Kenneth Bergeron's* Tritium on Ice. *It also references the 2003 article in the journal* Science and Global Security, *R. Alvarez et al., "Reducing the Hazards of Stored Spent Power-Reactor Fuel in The United States," which one of us (Lyman) co-authored. In addition, it draws on numerous NRC documents originally withheld from the public on the SOARCA program that were released to UCS under the Freedom of Information Act in 2011. It also relies on other NRC and ACRS public documents, including some released under Fukushima-related FOIAs.*

1. The Center for Responsive Politics, which tracks lobbying and campaign expenditures, reported that from 2008 through 2010 the Nuclear Energy Institute spent more than $6 million lobbying the federal government on nuclear issues. Large utilities and nuclear manufacturers spent millions more.

On February 9, 2012, the NRC commissioners voted 4–1 to approve licenses to construct the reactors at the Vogtle nuclear complex, about 170 miles east of Atlanta. (NRC chairman Gregory Jaczko cast the lone "no" vote, saying, "I cannot support issuing this license as if Fukushima had never happened." Jaczko thought that lessons learned from the Japanese disaster could lead the NRC to adopt new requirements for U.S. plants, and he argued that any new licenses should be conditioned on plant owners' agreement to incorporate all safety upgrades that the NRC might require in the future.) Two reactors already were operating at the facility, and upon completion this would become the nation's largest private multiple-reactor complex.

2. In 1983, for example, the NRC required that any testimony on the consequences of an accident at the Indian Point nuclear plant also had to address the probabilities. In all cases, the NRC argued, those were very low.

3. One of the authors of this book, Edwin Lyman, was among several co-authors of the study.

4. McGaffigan was cleared after the NRC staff members told the inspector general that they had arrived independently at their conclusions rebutting their study. The staff and the commissioner were conveniently on the same wavelength.

5. The NRC Office of Nuclear Regulatory Research conducts some of its technical work in-house but farms out most of it to external contractors, primarily staff at the Energy Department's national laboratories.

6. The two Peach Bottom reactors are almost identical to Fukushima Daiichi Units 2–5.

7. The NRC uses the term *conservative* in the sense of *conservative factor of safety*, a widely used term related to building in extra safety margins for various products or models.

8. The CRAC2 study used accident source terms that dated to the 1975 Reactor Safety Study. The worst-case source term was denoted "SST1." SST1 postulated a release, at ninety minutes after the accident began, of a very large fraction of the radioactive material within the core.

9. The CRAC2 radiation releases were generally much greater than the SOARCA values: 67 percent of the cesium-137 and 45 percent of the iodine-131 in the core, compared with 2 percent and 12 percent, for example. And the CRAC2 release was assumed to begin one and a half hours after the station blackout occurred, while the SOARCA release occurred eight hours for the Peach Bottom short-term station blackout.

10. When the SOARCA study was finally released in January 2012, the grand "communications plan" that had been such a priority when the project was launched years earlier included its own type of numerical sleight of hand: the size of the typeface. The brochure detailing the results featured bar graphs including only the data from the ten-mile zone in its comparison to CRAC2. For the areas beyond, small print on the next page noted that "the difference diminishes when considering larger areas."

11. 2012: "The Government Owes the Public a Clear and Convincing Answer"

Contemporaneous news accounts were used in this chapter.

1. In that regard, the limited, carefully scripted accident assumptions resembled the SOARCA study methodology used by the NRC, which disregarded accidents deemed too improbable and—at least in the mitigated models—assumed that all emergency actions were successful.

2. Meserve was likely reflecting on his own experience as NRC chairman. In late 2001, he presided over the agency when it made the near-disastrous decision to allow the Davis-Besse plant in Ohio to continue operating until the next scheduled refueling outage, despite accumulating evidence of a potentially serious problem at the plant's single reactor. Based on circumstantial but compelling evidence, the NRC staff had drafted an order that would require the plant to be shut down for safety inspections, a step the NRC had not taken since March 1987. But senior NRC managers shelved the order for economic reasons. When the reactor was finally shut down, workers discovered a pineapple-sized hole in the reactor vessel head. The plant had come within months of an accident that could have exceeded the severity of Three Mile Island.

3. In Japan, utilities set their own rates in a complex formula that rewards spending; the more a utility spends, the more it can charge.

4. The company also asked for and received approval for additional funds totaling about $30 billion (2.4 trillion yen) to compensate the victims of the nuclear accident. (It was the third time TEPCO had come back for more money to help victims.)

5. One bright spot was that even with the increase in the use of fossil fuels, Japan was still on track to meet its carbon emission reduction targets under the Kyoto Protocol.

6. The task force inadvertently sabotaged Recommendation 1 by its very wording: "The Task Force recommends establishing a logical, systematic, and coherent regulatory framework for adequate protection that appropriately balance defense-in-depth and risk considerations." A commissioner agreeing with Recommendation 1 could be viewed as implicitly conceding that the agency had long taken an illogical, chaotic, and incoherent approach to protecting American lives. Had the task force phrased this recommendation more adroitly, it might have fared better.

7. Here's a simple illustration of this point. Current probability risk assessments do not take into account the possibility that an accident at one reactor will trigger an accident in an adjacent unit. So a safety improvement that would reduce the risk of this happening by 50 percent—say, by controlling hydrogen explosions in the reactor building of a Mark I BWR so that debris could not damage safety equipment at a neighboring reactor—would register as having zero benefit in the cost-benefit calculation. There would be no benefit because the risk wasn't included in the first place: 50 percent of zero is still zero.

8. Spent fuel pools typically do not have any drains or piping connections below the normal surface of the water to protect against inadvertent drainage. Water in a spent fuel pool flows into scuppers—similar to those in many swimming pools—that channel it into collection tanks. Pumps draw water from the collection tanks and route

it through heat exchangers and filter demineralizer units. The cooled and cleaned water is poured back into the pools from above. A common arrangement monitors the level and temperature of the water in the collection tank—not in the pool itself. When the cooling system is operating, the tank and pool conditions match. But if cooling is impaired or lost, the tank and pool conditions can be vastly different. The NRC's order required operators to provide a reliable and accurate means to monitor the level of the water inside the spent fuel pool under certain accident conditions.

9. The task force timelines themselves did not take into account what actually happened at Fukushima Daiichi. There it took workers fifteen hours to begin emergency coolant injection into the Unit 1 core, by which time the core had already melted. Even if the installed coolant systems had worked for eight hours, they would not have sufficed to prevent a meltdown. In short, if the order had been in place at Fukushima, it would have provided no guarantee that disaster could have been averted.

10. The call to phase out nuclear power placed some of Japan's largest industries in an awkward position. Mitsubishi Heavy Industries, Toshiba, and Hitachi are among the world's leading manufacturers of nuclear power components. Marketing technology abroad that was deemed unacceptably risky at home might have proven difficult. The U.S. nuclear industry also had a vested interest in the outcome of the phaseout debate. Since 2007, General Electric and Hitachi have been business partners. In 2006, Toshiba purchased Westinghouse Electric Company. Japanese media carried reports that additional, unidentified U.S. interests were lobbying to influence Tokyo to retain its nuclear generating commitment.

12. A Rapidly Closing Window of Opportunity

Documents utilized in this chapter include Forging a New Nuclear Safety Construct *prepared for the American Society of Mechanical Engineers Presidential Task Force on Response to Japan Nuclear Power Plant Events (New York: American Society of Mechanical Engineers, 2012), files.asme.org/asmeorg/Publications/32419.pdf; the NRC's "Briefing on the Task Force Review of NRC Processes and Regulations Following the Events in Japan: Transcript of Proceedings," July 28, 2011, www.nrc.gov/japan/20110728.pdf; as well as hearing transcripts from the Senate Committee on Energy and Natural Resources, Subject S. 512, the Nuclear Power 2021 Act, June 7, 2011, www.energy.senate.gov/public/index .cfm/hearings-and-business-meetings?ID=237c8727-802a-23ad-41f3-e4cfc52bc3a2.*

1. Senator Franken could not be faulted for being unaware of this danger, for the NRC, citing security concerns, had concealed the agency's growing worry about the threat from public view for more than a decade.

2. For example, the industry adopted, and the NRC approved, this less-than-helpful standard for protection of FLEX equipment from high temperatures: "the equipment should be maintained at a temperature within a range to ensure its likely function when called upon."

3. In its comprehensive 2102 report on the Fukushima accident, the American Society of Mechanical Engineers, a professional engineering society with more than 127,000 members in 140 countries, noted that protecting the public from the health effects of radiological releases "continues to be the primary focus of nuclear safety." In

the case of Fukushima, radiation releases were low and not believed to pose a significant health threat. But the report identified long-lasting harm of another type: "The major consequences of severe accidents at nuclear plants have been socio-political and economic disruptions inflicting enormous cost to society." Those costs, long overlooked, now also must be factored in to risk/benefit equations, the ASME concluded.

Appendix

Material for this section was drawn primarily from oral presentations and the proceedings of the American Nuclear Society's International Meeting on Severe Accident Assessment and Management: Lessons Learned from Fukushima Dai-ichi, San Diego, California, November 11–15, 2012.

INDEX

Page numbers in *italic* refer to illustrations.

Abe, Shinzo, 50, 243, 277
accident management (AM) measures, 16, 202–3, 235, 245
accident simulation. *See* computer simulation and models
Act on Special Measures Concerning Nuclear Emergency Preparedness (Japan), 16, 38
acute radiation syndrome, 27, 190, 192, 208, 215, 219
"adequate protection," 187, 188, 190, 194–95, 197, 234, 238, 257, 261; NTTF on, 252, 253, 260
Advisory Committee on Reactor Safeguards, 153, 197–98, 214, 218, 269
aerial monitoring and mitigation, 69, 84, 90, 95–96, 97, 123
agriculture. *See* farmers and farming
Alaska, 99, 113, 114, 128, 129, 138
Amano, Yukiya, 105, 277
Ambassador Roos. *See* Roos, John V.
American Embassy, Tokyo. *See* U.S. Embassy, Tokyo
American Nuclear Society, 2012 meeting, 263–65
American Physical Society, 190, 205–6
Americans: helpful hints, 101–2; in Japan, 62, 68–69, 84–85, 87–89, 92–93, 108, 129, 132, 134, 138–39, 282n4; opinion of nuclear power, 146. *See also* evacuation of Americans; travelers, American
American Society of Mechanical Engineers, 290–91n3
americium-241, 128
antinuclear movement, 110, 146, 161, 205, 209–10, 211, 228

Aoki, Steven, 138, 277
Apostolakis, George, 182, *183*, 277
Areva, 102, 164, 245
Asahi Shimbun, 35, 47, 228, 229, 240
Asseltine, James K., 193–94, 201, 278
Atomic Energy Act, 187, 193, 194
Atomic Energy Commission (AEC), 39, 153, 187, 192, 273
Atomic Energy Commission of Japan, 77, 101
Atomic Industrial Forum, 189
attitudes, complacent. *See* complacency and overconfidence
auxiliary systems. *See* backup systems

Babcock & Wilcox, 142
backfitting, 15, 191–200, 234, 241, 254, 255, 259, 268; definition, 287n2
backup generators, 8, 10, 12, 53, 167, 271; in FLEX plan, 173, 237, 238; malfunction/inoperability, 13, 22, 24, 42, 250–51, 282n6; U.S. use, 175–76, 217
backup systems, 8, 10, 175–76, 214, 255; battery-operated, 12, 13, 17, 18, 65, 66; destroyed by tsunami, 250; failure, 13, 65, 188, 218–19, 248. *See also* backup generators; B.5.b equipment/measures; "defense-in-depth"; emergency cooling systems; FLEX program; isolation condensers
bailouts, 178–79, 228 , 289n4
Barrasso, John, 182, 185
batteries, 12, 18, 19, 30, 31, 67, 74; assumptions about, 237, 251; failure, 13, 65, 218; NRC views, 167

Beasley, Benjamin, 116
Bechtel Corporation, 98, 100
Beck, Glenn, 102
Bernero, Robert, 190, 208
"best-estimate" scenarios, 130, 139, 212, 213
beyond-design-basis accidents, 14, 153, 167, 169, 173, 188, 234, 235, 237, 269; assumptions about, viii, 16, 140; NRC views, 195, 202, 206, 208, 253; planning for, 20, 169–70, 188–91, 199–203, 218, 236, 237, 252, 253; Virginia, 175–76. *See also* Chernobyl nuclear accident, 1986; severe accident management guidelines (SAMGs); severe accident mitigation alternatives (SAMA); Three Mile Island nuclear accident, 1979; Tokaimura research center: nuclear accident, 1999
B.5.b equipment/measures, 95, 169, 173, 215, 218, 235, 269, 285n4
blackouts. *See* power outages
boiling water reactors (BWR), 5–6, *6*, 70–71, 77, 84, 245, 269. *See also* cooling systems; Mark I containment; Mark II containment; Mark III containment
Bonaccorso, Amy, 101–2
Borchardt, Bill, 65, *86*, 89–90, 91, 94, 139, 232–33, 278
boron and boric acid, 62, 88, 98, 283n2
Boxer, Barbara, 94, 95, 182
Bradford, Peter, 258
breeder reactors, 171, 286
Brenner, Eliot, 114, 116, 278
Bromet, Evelyn, 111
Browns Ferry Nuclear Power Plant, 248, 249, 250, 261
Burnell, Scott, 112, 116, 278
Burns, Shawn, 217–18

California, 94, 134, 139, 164, 182. *See also* Diablo Canyon Power Station; San Onofre Nuclear Generating Station
cameras, remote, 12, 32, 90, 97, 266
cancer, 27–28, 84, 108; AEC risk study, 192; CRAC2 quantification, 207, 208, 216, 219; IDCOR projections, 190; NUREG-1738 study, 211; SOARCA projections, 213, 216, 219, 220; Yoshida's, 284n1

Candris, Aris, 245, 246
Carney, Jay, 92
Casto, Charles "Chuck", 86–90, *86*, 93, 95–101, 124–25, 131–35, 155, 163, 278; attire, 284–85n2; shortchanged by Office of Research, 285n3; speech at RIC, 260–61
casualties. *See* fatalities
Center for Responsive Politics, 287n1
cesium, 60, 62, 82, 118, 146, 155, 156, *156*, 268; in food, 178
cesium-134, 159
cesium-137, 71, 97, 126–27, 157, 159, 164, 211, 267, 288n9; aerial measuring, 97; in food, 157
Chernobyl nuclear accident, 1986, 20, 44, 45–46, 100, 104, 130, 164, 202, 209, 249, 261; casualties, 27, 90; considered irrelevant, 201, 205; Fukushima compared, 268, 272; INES rating, 104; mental health aspects, 111; mitigation, 90, 100; permanent exclusion zone, 178; population resettlement, 164–65
children, 58, 89, 109, *109*, 139, *161*, 162, 178, 179, 223; Alaska, 129; California and West Coast (U.S.), 134, 139; cancer risk, 28; evacuees, 120, 148, 160, 241, 242
China, 25, 38, 44, 100, 124, 160, 171
Chubu Electric, 41, 170
Citizen Scientist (von Hippel), 192
Citizens' Nuclear Information Center, 106
cladding, 5, 6, 7, 57, 66, 127; ignition of, 57, 71, 82, 91, 113, 127, 145; Three Mile Island, 14, 145
classified information, 209, 211
cleanup, 157, 159, 163, 164, 177–78, *179*; by citizens, 162; in CRAC 2 modeling, 207; Three Mile Island, 146, 164
cleanup costs, 84, 165, 177–78, 179, 214, 226, 254, 289n4; ASME on, 290–91n3; CRAC2 model, 207; Three Mile Island, 146, 164
cobalt, 155
cold shutdown, 62, 163, 180, 222, 269
Cold War, 38, 39, 207
common-mode failure, 250, 269
communications, 16, 17, 22, 25, 26, 76–77, *86*, 87; bungled, 21, 76; e-mail,

35–36, 46, 54, 101, 112, 115–16, 185, 217, 220; intra-industry, 151; Japanese government, 23, 26, 59, 103, 104–5, 108–9, 157; phone, 17, 148; Three Mile Island, 145, 147, 148; video, 25, 86. *See also* information sharing; information withholding; media; press conferences; press releases; public relations; secrecy

company uniforms, *23*, 67, 285n1

compensation to victims, 165, 178–79, 226

complacency and overconfidence: fostered by defense-in-depth approach, 250; Franken's, 247–48; Japanese, 12, 16, 44, 225, 226; noted after Three Mile Island accident, 150; NRC, 167, 168, 185–86, 200–201, 203, 208, 216, 221; in plant design and siting, 51; in reactor design, 70

computer simulation and modeling, 14, 79, 125, 136, 218; Casto view, 87–88; Exelon, 135; failure to explain Fukushima events, viii, 263, 264; NARAC, 126; post–Three Mile Island, 189; of radioactive exposure, 99; for SOARCA, 212, 215; by TEPCO, 179. *See also* CRAC2 study; MELCOR; RASCAL (Radiological Assessment System for Consequence Analysis)

concrete, 5–6, 7, 14, 70, 75, 113, 122, *123*, 136, 145–46, 180, 188; reaction with molten fuel, 7, 179–80, 264; in storage casks, 7, 83

conflict of interest, 46

Congress. See U.S. Congress

consumer electricity rates. *See* electricity rates

contaminated food. *See* food contamination

contaminated homes, 206, 207, 220, 230. *See also* evacuation of residents

contaminated materials, disposal and storage of. *See* disposal and storage of contaminated materials

contaminated soil. *See* soil contamination

contaminated water. *See* radioactive water

contamination levels. *See* radiation levels

control rods, 5, 49, 269–70, 282n2, 283n2

control rooms, 5, 8, 49, 142, 150; blackouts, 13, 17–20, 25, 74, 124; Browns Ferry, 248; communications, 9, 17, 147–48; evacuation, 75, 85; radiation levels, 29, 85; remote operations from, 19–20, 25, 31, 32, 66, 241, 284n7; secondary, 241; stress level in, 144; Three Mile Island, 144, 147–48, 157; Units 1 and 2, 13, 17, 18, 19, 22, 25, 29, *29*, 31, 32, 124; Units 3 and 4, 17–18, 66, 67, 73

cooling systems, 7–8, 12, 70, 84, 94, 98, 142, 164, 184, 218, 248, 250; effect on design, 51; Three Mile Island, 142–47. *See also* emergency cooling; spent fuel pools

"core catchers," 188, 245

core meltdowns. *See* meltdowns

corruption and collusion, 46, 47, 48

cost-benefit analysis, 191–92, 193, 195, 196, 234, 256, 259, 268, 289n7; definition, 270; Fitzpatrick nuclear plant, 198; of "recombiners," 202; terrorist attacks excluded from, 209

cost of cleanup. *See* cleanup costs

cost of human life. *See* fatalities; human life, monetary value

coveralls and uniforms, *23*, 67, 285n1. *See also* protective clothing

cover-ups, 48, 184, 206–8

CRAC2 study, 207–9, 211, 212, 214, 215, 216, 219, 220

Cuomo, Andrew, 116

Curran, Diane, 194

Curtiss, James, 198

dairy industry. *See* milk contamination

Daley, William, 177

dam failure, 183–85, 248

Davis-Besse Nuclear Power Station, 143–44, 151, 289n2

deaths. *See* fatalities

DC power, 8, 32, 167, 237, 282n3; definition, 270; loss of, 13, 17, 26, 167, 202. *See also* batteries

decay heat, 7, 65, 189, 270

decontamination. *See* cleanup

Dedman, Bill, 115–17, 278

Deepwater Horizon oil spill, 2010, 102

Defense Department. *See* U.S. Department of Defense

"defense-in-depth," 196, 248–51, 252, 253, 254, 255, 257, 270, 289n6
Defense Threat Reduction Agency, 97
demonstrations. *See* protests
Denton, Harold, 101–2, 148, *149*
Department of Defense. *See* U.S. Department of Defense
Department of Energy (DOE). *See* U.S. Department of Energy (DOE)
Department of Homeland Security. *See* U.S. Department of Homeland Security
Department of State. *See* U.S. Department of State
design-basis, 169, 173, 184, 270
design-basis accidents, 13, 14, 115, 144, 153, 187, 188
Diablo Canyon Power Station, 34, 98, 115, 128
diesel generators, 8, 10, 53, 167, 168, 169, 214, 250, 271; confidence in, 12, 168, 282n6; malfunction/inoperability, 10, 22, 250–51, 282n6; U.S. emergency use, 175–76, 249
diesel pumps, 19, 66, 95, 169, 214
direct current. *See* DC power
dirt, 91, 160, 162. *See also* soil contamination
disasters. *See* earthquakes; floods; tsunamis
disease, 27–28. *See also* cancer
disposal and storage of contaminated materials, 163, 164, 243
DNA damage, 27, 28
Dominion, 176
Donald, Kirkland, 68, 89, 278
Dorman, Daniel, 54, 87, 99, 278
dose rates. *See* radiation levels
dosimeters, 22, 59, 85, 270
drills, emergency. *See* emergency exercises
drone use, 69, 97
dry cask storage, 83–4, 211, 256, 270
drywell, 5–6, 55, 270–71; breach of, 26, 28, 57, 264; Unit 1, 25, 26, 28, 57; Unit 2, 74, 267; Unit 3, 69
Duke Energy, 185
Dyer, Jim, 139

Earthquake Countermeasures Act (Japan). *See* Large-Scale Earthquake Countermeasures Act (Japan)

earthquakes, 1–5, 9–10, 41, 47–48, 51, 94, 95, 246; aftershocks, 29, 30, 73, 85; assumptions about, 53, 213, 214; consequences foreseen, 211, 217–18, 245; early-warning systems/detection, 1, 3–4, 114–15; effect on pipes, seals, etc., 159; forecasting, 42–43, 44, 52, 229; Franken joke about, 247; magnitude scales, 281–82n1; Niigata, 2007, 19, 42, 50, 225; planning for, 63, 169, 170–71, 201; Sanriku, 1896, 9–10, 52, 54; shutdown after, 5, 114–15; spent fuel pools and, 71; United States, 113, 114, 175–76, 249–50, 253, 287n6. *See also* Jogan earthquake; Tohoku Earthquake, 2011
economic costs. *See* bailouts; cleanup costs
The Economist, 178–79, 227
Edano, Yukio, 23, 26, 59, 96, 103, 104, *104,* 108, 109, 165, 277; downplays food contamination, 157; as minister of Economy, Trade and Industry, 179, 226–27, 231, 240; on-camera dress, 285n1,
Education Ministry. *See* Ministry of Education, Culture, Sports, Science and Technology (MEXT)
elections, 15, 240, 243
electrical cables, 13, 19, 24, 57, 124, 158, 248
electrical grid incompatibility, 282–83n8
electrical outages. *See* power outages
electricity use and supply (electrical grid), 38, 171–72, 205, 224, 227, 228–29; rates, 179, 227, 289n3. *See also* electrical grid incompatibility
Electric Power Research Institute, 102
emergency backup systems. *See* backup systems
emergency cooling, 8, 13, 14, 33, 49, 69, *72,* 93, 131, 167, 251, 290n9; "feed and bleed," 158, 163, 271; FLEX plan, 173, 174, 254, 256; in new reactor designs, 244, 245; policy recommendations, 168–69; Unit 1, 17, 19, 25, 33, 60–61, 122; Unit 2, 17, 97, 122; Unit 3, 65, 66, *122;* U.S. involvement/proposals, 98, 100, 101, 189, 197, 199, 255–56;

vulnerability, 180. *See also* emergency core cooling systems; fire engines, pumps, etc.; seawater: emergency use
emergency core cooling systems, 8, 17, 187, 241, 271; Browns Ferry, 248; failure/loss of, 18, 49. *See also* high-pressure coolant injection system (HPCI); reactor core isolation cooling systems (RCIC)
emergency evacuation. *See* evacuation of Americans; evacuation of residents; evacuation of workers
emergency exercises, 25, 62, 78, 142, 152, 287n4
emergency generators. *See* backup generators
emergency planning zone, 129, 140, 215–16, 221, 256, 271
emergency procedures manuals. *See* manuals
emergency systems. *See* backup systems
emergency workers, 22, 85, 93, 106, 181, 243; vulnerability, 94, 158, 284n8
energy consumption. *See* electricity use and supply
Energy Department. *See* U.S. Department of Energy (DOE)
energy policy, Japanese, 47, 108, 165, 171–72, 222–23, 227, 231, 239–40
Energy Policy Act (EPAct), 204–5
Environmental Protection Agency. *See* U.S. Environmental Protection Agency (EPA)
equipment, backup. *See* backup systems
equipment upgrades. *See* retrofitting
European Union, 224. *See also* Finland; France; Germany; Sweden
evacuation of Americans, 84–85, 87, 88–89, 92, 108, 129, 132, 134, 138–39
evacuation of residents, 16, 24, 30, 31, 37, 57–59, 77, 117–20, *117*, 160–61, 241–43; 285n2; Chernobyl, 46; compensation for, 165, 178–79, 207, 226, 289n4; computer modeling, 80, 87, 207, 215–16; housing and services, 117, 157, 160, 178; long-term status, 230; simulations/exercises, 152; statistics, 45; Three Mile Island, 141, 148–49. *See also* evacuation zones
evacuation of workers, 12, 76

evacuation zones, 77, 108, 117, 120, 140, 159, 241, 256; initial expansion of, 30; maps, *119, 242*; media steers clear of, 105–6; NRC recommendations, 80, 87, 90, 92–93, 99, 139–40; radiation levels, 178; Three Mile Island, 148; U.S. regulations, 190
Exelon Corporation, 61, 102, 135, 174, 236–39
exhaust stacks, *4*, 32, 57, 67, 266
explosions: Chernobyl, 46; effect on pipes, seals, etc., 159; Unit 1, *32*, 54, 55–57, *56*, 61–62, 71, 93, 122; Unit 2, 74–75, 78, 98; Unit 3, 72, *72*, 81, 118, 265; Unit 4, 75–76, *75*, 80–82, 88, 91. *See also* hydrogen explosions
exposure to radiation. *See* radiation exposure

failure, common-mode. *See* common-mode failure
fallout, 99, 114, 118, 156, 159. *See also* cleanup
falsification of reports, 46, 48–49, 50
farmers and farming, 119, 120, 157
fatalities: cattle, 120; Chernobyl, 27, 90; CRAC2 projections, 207, 208, 209, 215; after earthquake/tsunami, 9, 10, 43, 118, 181; Hiroshima and Nagasaki bombings, 39; "latent" (from cancer, etc.), 28, 84, 190, 192, 207, 208, 211, 213, 216, 219, 220, 256; NUREG-1738 projections, 211; risk assessment, 192; Tokaimura, 37
fault lines, 41, 42, 44, 50, 52, 170, 229
fears, public. *See* public fears
Federal Emergency Management Agency, 152
Fertel, Marvin, 232
filters and filtering, 17, 28, 267; in cleanup, 164; ventilation, 20, 128, 188, 231, 241, 254–55, 268; of water, 60, 289n8; water used for, 28, 74
Finland, 189
fire engines, pumps, etc., 19, 77, 100, 124, 131, 202, 265; damage to, 72; inadequacy, 29, 30, 97; Unit 1, 19, 30, 33, 57, 62, 65; Unit 2, 73, 74; Unit 3, 66, 69, 96, 97, 121, *122*; Unit 4, 136; U.S., 97

fires, 50, 71, 87, 91, 211; Browns Ferry, 248, 250; possibility/threat of, 167, 173, 217, 248; protection from, 239; Unit 3, 99; Unit 4, 81–82, 133

fishing industry, 156

fission, 5, 7, 46, 269–70

fission products, 7, 57, 66, 70, 109, 126–27, 187, 189, 267

FitzPatrick nuclear plant. *See* James A. FitzPatrick Nuclear Power Plant

FLEX program, 173–74, 236–39, 254–55, 256, 259, 271, 290n2

flooding, 167, 169, 238; assumptions about, 214, 217–18, 238; in computer models, 219; after dam failures, 183, 184, 248; Fukushima, 10, 13, 124, 137, 141, 159; United States, *174,* 175, 176, 249

floodwalls, 185. *See also* seawalls

Florida, 249

food contamination, 127, 129, 139, 156, 157, 178, 207

foreign relations, Japanese, 125, 131, 160

Fort Calhoun Nuclear Generation Station, *174,* 175, 176, 249, 250

France, 102, 125, 164, 245, 247

Franken, Al, 247, 290n1

freshwater, 286n2; emergency use, 30, 33, 57, 67, 135, 137; normal use, 60, 131

fuel assemblies, 5, 48, 57, 70, 83, 84, 266, 271

Fukui Prefecture, 224, 230

Fukushima Daiichi design and construction, 5–6, 7, 8, 51, 52, 53–54. *See also* Seismic Isolation Building; Mark I containment; spent fuel pools

Fukushima Daini nuclear plant, 18, 40, 282n7

Fukushima Prefecture, 10, 16, 24, 40, 52, 105, 156, 157, 178, 181, 253; contamination levels, 163; schools, 160, *161,* 162

Futaba, Japan, 16, 24, 40, 118

gamma rays, 127

gas, natural. *See* natural gas

gases. *See* hydrogen; nitrogen; radioactive gases

gauges, 19, 23–24, 65, 74, 81, 150;

calibration insufficiency, 263–64; jury-rigged, 55; lacking, 144

GE Hitachi Nuclear Energy, 87, 135, 290; ESBWR, 114, 245

General Electric Corporation (GE), 5, 14, 39, 40, 87, 244, 245, 290n10; collaboration, 102, 132; employees, 48. *See also* Mark I containment; Mark II containment; Mark III containment. *See also* GE Hitachi Nuclear Energy

General Public Utilities Corporation, 148

generators. *See* backup generators; diesel generators; portable generators; turbine generators

Germany, 124, 282n8

golf courses, 179

Goto, Masashi, 225

government bailouts. *See* bailouts

Great Tohoku Earthquake. *See* Tohoku Earthquake, 2011

ground motion, 1, 3, 44, 50, 114–15, 175

Grobe, Jack, 77–78, 84–85, 278

Hachiro, Yoshio, 179

half-life, 127

Hamaoka nuclear plant, 41–42, 44, 170, 229

hardened vents, 63, 203, 268, 272, 283n4; incorporated at Fukushima, 169, 202; NRC views and actions, 197, 198–99, 235, 254

Hardies, Robert, 54

Harrisburg, Pennsylvania, 141, 146, 147, 148

Hashimoto, Toru, 230, 231, 277

Hatamura, Yotaro, 108, 226, 277

hazmat suits. *See* protective clothing

Headquarters for Earthquake Research Promotion, 44, 52

health consequences, 27–28, 108, 127, 158, 219; mental health, 111; NRC views, 205, 211, 213, 216, 219; projected by CRAC2, 206, 207. *See also* cancer

Hearn, Lafcadio, 282n4

heat sink, 7, 236, 272

Hein, Laura E., 38

helicopters, 1, 21, 30; mitigation from, 83, 90, 95, 96–97, 99, 121, 123–24, 136, 137; radiation on, 68

Hendrie, Joseph, 145, 187
high-pressure coolant injection system
  (HPCI), 65–66, 69, 264, 265, 272
Hirose, Naomi, 228, 277
Hiroshima bombing, 1945, 27–28, 49
Hitachi-GE partnership. *See* GE Hitachi
  Nuclear Energy
Holahan, Trish, 129, 130–31
Homeland Security Department. *See* U.S.
  Department of Homeland Security
homes, contamination of. *See*
  contaminated homes
Hosono, Goshi, 177–78, 277
HPCI. *See* high-pressure coolant injection
  system (HPCI)
Hudson Riverkeeper, 209
human error, 94, 133–34, 140, 142–43, 225;
  Futaba, 118; Sandia view, 217; Three
  Mile Island, 45, 142–43, 149–50, 188
human life, monetary value, 193
hurricanes, 94, 249, 253
hydrogen, 7, 28, 57, 66, 74, 91, 145, 188,
  200, 264, 265, 287n5
hydrogen explosions, 14–15, 56, 61–62,
  71, 75–76, 80–82, 91, 122, 219, 266;
  attempts to explain, 264; positive side
  effects, 84, 266; prevention, 15, 55, 195,
  200, 202; Three Mile Island, 141, 145,
  188, 195

IAEA. *See* International Atomic Energy
  Agency (IAEA)
IDCOR. *See* Industrial Degraded Core
  Rulemaking program (IDCOR)
Iitate, Japan, 118–20, 160, 267
imported fuel reliance, 171, 228–29
Indian Ocean earthquake, 2004, 44, 47
Indian Point Energy Center, 115–16, 208,
  209, 214, 287n6, 288n2
Individual Plant Examination (IPE)
  program, 191, 199
Industry Degraded Core Rulemaking
  program (IDCOR), 189–90, 196, 205
industry measures, voluntary. *See*
  voluntary industry measures
"inerting," 15, 55, 195, 200, 202
INES. *See* International Nuclear and
  Radiological Event Scale (INES)
information sharing, 35, 62, 69, 87, 96,

151, 155. *See also* public disclosure and
  participation
information withholding, 58, 107, 108,
  109, 118, 151–52, 216. *See also* classified
  information; cover-ups; secrecy
Ino, Hiromitsu, 225
inspections. *See* nuclear plant inspections
Institute of Nuclear Power Operations
  (INPO), 61, 102, 131, 151–52
instrument scale problems, 144–45, 158,
  263–64
insubordination, 60–61, 86
International Atomic Energy Agency
  (IAEA), 22, 50, 105, 119, 224–25, 228,
  268
International Commission on Radiological
  Protection, 161
International Nuclear and Radiological
  Event Scale (INES), 22, 103–4, 109–10,
  272
Internet, 12, 35, 106–7, 109, 110
Investigation Committee on the Accident
  at the Fukushima Nuclear Power
  Stations, 108, 226
iodine, 62, 118, 125, 140, 146, 155, 157,
  256. *See also* potassium iodide
iodine-131, 71, 126–27, 133, 146, 267–68,
  288n9; in California, 128; in milk, 157;
  in seawater, 156; Unit 2, 159, 267; Unit
  3, 158; Unit 4, 133
IPE program. *See* Individual Plant
  Examination (IPE) program
irradiated fuel. *See* spent nuclear fuel
Ishibashi, Katsuhiko, 41, 42, 48, 277
isolation condensers, 8, 13, 17, 18, 22, 25,
  264, 282n3

Jaczko, Gregory: adherence to script,
  114; at congressional hearings, *92,* 182,
  183; event monitoring (days 2–4), 54,
  61, 62–64, 77; event monitoring (days
  5–8), 79, 84–85, 89–90, 92, 93–94, 95,
  96, 113, 114; event monitoring (days
  9–10), 130–31, 138; management style
  and resignation, 176–77, 233; mandate
  for recovery, 166; at odds with fellow
  commissioners and NRC staff, 172–73,
  176–77, 212, 232–33, 245, 288n1; on
  worst-case scenarios, 130

James A. FitzPatrick nuclear plant, 198–99
Japan Atomic Energy Commission. *See* Atomic Energy Commission of Japan
Japan Defense Ministry. *See* Ministry of Defense
Japan Education Ministry. *See* Ministry of Education, Culture, Sports, Science and Technology (MEXT)
Japanese Diet Independent Investigation Commission, 244–45
Japanese Embassy, Washington, DC, 131
Japanese energy policy. *See* energy policy, Japanese
Japanese foreign relations. *See* foreign relations, Japanese
Japanese nuclear plant licensing. *See* licensing of Japanese nuclear plants
Japanese nuclear plants. *See* Fukushima Daini nuclear plant; Hamaoka nuclear plant; Kashiwazaki-Kariwa nuclear plant; Monju nuclear plant; Ohi nuclear plant
Japan Meteorological Agency, 1, 3, 4, 15, 281n1
Japan Ministry of Defense. *See* Ministry of Defense
Japan Ministry of Economy, Trade and Industry (METI). *See* Ministry of Economy, Trade and Industry (METI)
Japan Ministry of Education. *See* Ministry of Education, Culture, Sports, Science and Technology (MEXT)
Japan Nuclear Energy Safety Organization, 16
Japan Nuclear Safety Commission. *See* Nuclear Safety Commission (NSC)
Japan Science and Technology Agency, 37, 38
Japan Self-Defense Forces. *See* Self-Defense Forces (SDF)
Japan Society of Civil Engineers, 52
Japan Times, 171
JCO Company, 37
Jocassee Dam, *184*, 185
Jogan earthquake, 9–10, 52, 54
Justification for Continued Operation, 185

Kaieda, Banri, 26, 30, 277
Kammerer, Annie, 115, 116, 278

Kan, Naoto, 15, 225, 277; calls for nuclear phaseout, 171–72; disaster response, 17, 22–26, *23*, 30–31, 37, 55, 59–61; during earthquake, 1, 15; information withheld from, 58; NSC relations, 109; orders to residents, 30, 76; post-disaster, 23, 26, 170–72, 224, 240, 285n5; Shimizu relations, 67, 76–77, 243; U.S. relations, 95, 96
Kansai Electric Power Company, 230
Kashiwazaki-Kariwa nuclear plant, 18, 19, 42, 49, 50, 225, 228, 239, 283n2
Katsumata, Tsunehisa, 21, 50, 226, 277
Keidanren, 240
Kemeny, John G., 150
Kemeny Commission, 147, 149, 150, 186
Kennedy, Robert F., Jr., 209
Kobe earthquake, 1995, 3, 44, 47
Kosako, Toshiso, 162, 277
krypton-85, 146
Kurion, 164
Kyoto Protocol, 289n5

land contamination. *See* soil contamination
Large-Scale Earthquake Countermeasures Act (Japan), 42, 43
lawsuits, 41–42, 179, 194
liability protection, 246
licensing of Japanese nuclear plants, 42, 44
licensing of U.S. nuclear plants, 111, 113–14, 170, 175, 184, 205; amendments, 198, 287n3; Fort Calhoun, 175; renewal, 259; Vogtle, 245, 288n1
Limerick Nuclear Power Plant, 214
linear no-threshold hypothesis (LNT), 216
lobbying, 36, 94, 287n1, 290n10
Lochbaum, David, 201, 283n1
Luxat, David, 264
Lyman, Edwin, 94, 95, 288n3

Macfarlane, Allison, 176, 177, 278
Madarame, Haruki, 31, 41–42, 47, 55, 59, 60, 109, 225, 277
Magwood, William, 182, *183*, 278
manual operations, 20, 25, 28–29, 31, 66, 201, 251, 284n7
manuals, 20, 251
maps, *2, 45,* 98, *119,* 126, *242*

Markey, Edward, 206, 208, 212, 214
Mark I boiling water reactor, 40, 204
Mark I containment, 5–6, 6, 15, 63, 70–71, 87–88, 186, 195–200, 244, 289n7; dilemmas, 195; at Peach Bottom, 213, 237; retrofitting, 202, 235, 254; ventilation, 19–20, 169, 235, 268
Mark II containment, 15, 70–71, 169, 200; retrofitting, 202, 235, 254, 268; ventilation, 235, 268
Mark III containment, 200, 287n5
McDermott, Brian, 64, 79
McGaffigan, Edward, 209–10, 210, 211, 217
media, 26, 34, 35, 58, 106, 112; antinuclear views, 171; Japan, 1, 46, 47, 58, 59, 67, 105–7, 110, 111, 171, 229, 241; McGaffigan view of, 210; NRC reliance on, 54, 61, 84, 85, 136; United States, 64, 85, 111–16, 119, 146–48, 180–81, 214, 286n1; on contamination, 178. See also press conferences; press releases; television
Mehta, Zubin, 100
MELCOR, 139, 140, 264, 272, 286n3
meltdown, 7, 12, 78, 79, 82, 89, 179–80, 264, 290n9; assumptions about, 130, 132, 187–88; design anticipating, 188, 245; fears of, 136; government belated acknowledgment of, 109–10; not considered in planning, 71, 239; NRC analysis, 195–96; NRC public relations view, 113; prevention, 20, 153, 197, 218; radionuclide release during, 127–28; terrorist-caused (hypothetical), 209, 211; Three Mile Island, 45, 141, 145, 147; Unit 1, 22, 24, 30, 55, 59–60, 62, 64; Unit 2, 78; Unit 3, 66; Unit 4, 82
meltdown (word), 60, 272
mental health, 111
Meserve, Richard, 226, 289n2
Metropolitan Edison, 142, 147–48
micromanagement, 25, 212
Middletown, Pennsylvania, 146–47
military. See Self-Defense Forces (SDF); U.S. military
milk contamination, 129, 139, 157
Miller, Charles R. "Charlie", 64, 90, 138, 139, 169, 278

Milne, John, 42–43
Minamisoma, Japan, 106–7
Ministry of Defense, 132
Ministry of Economy, Trade and Industry (METI), 15, 16, 22, 26, 30, 100, 110, 161
Ministry of Education, Culture, Sports, Science and Technology (MEXT), 15, 16, 58–59, 178
Minnesota, 247–48
Missouri River flooding, 2011, 174, 175, 176
modular reactors, 205, 245–46
Mogi, Kiyoo, 43–44, 277
moment magnitude scale, 281–82n1
Monju nuclear plant, 286n1
Monninger, John, 88, 90–91, 95, 100, 125, 135, 136–37, 278
monopolies, 38, 227
Mullen, Mike, 129–30, 278
Murray, Thomas E., 39

Nagasaki bombing, 1945, 27–28, 39
Nakamura, Koichiro, 59, 277
Namie, Japan, 57–58, 59, 118
National Atmospheric Release Advisory Center (NARAC), 126, 129, 130, 132, 133–34, 137–38, 139, 268
National Environmental Policy Act, 259
National Geographic, 282n4
National Nuclear Security Administration, 68, 69, 97
National Oceanic and Atmospheric Administration, 34
National Seismic Hazards Maps, 44
natural disasters. See earthquakes; floods; tsunamis
natural gas, 205, 228–29
Nature, 38
Naval Reactors. See U.S. Department of Energy (DOE): Office of Naval Reactors
NBC News, 115–17
NEI. See Nuclear Energy Institute (NEI)
new media, 106–7
news conferences. See press conferences
news media. See media
New York nuclear plants, 40, 115–16, 190, 198–99, 208, 209, 214, 287n6, 288n2

New York State Power Authority, 198

NHK (television network), 1, 12, *32*, 97

Niigata earthquake, 2007, 19, 42, 50, 225

9/11 terrorist attacks. *See* September 11, 2001, terrorist attacks (9/11)

Nine Mile Point nuclear plant, 40

NISA. *See* Nuclear and Industrial Safety Agency (NISA)

Nishizawa, Toshio, 166, 277

nitrogen, 15, 55, 195, 200, 202

Noda, Yoshihiko, 172, 179, 180, 222–23, 224, 229, 231, 240, 243, 277

North Anna Power Station, 175–76, 249–50

NRC. *See* Nuclear Regulatory Commission (NRC)

Nuclear Accident Independent Investigation Commission, 111, 277

nuclear accidents. *See* design-basis accidents; beyond-design-basis accidents; Chernobyl nuclear accident; Three Mile Island nuclear accident; Tokaimura research center: nuclear accident

Nuclear and Industrial Safety Agency (NISA), 15, 16–17, 22, 38, 49, 82, 229, 241; communications, 25, 26, 58, 59–60, 100, 103; encourages use of shills, 46; hearings on earthquake and tsunami hazards, 53–54; INES ranking by, 103; protests against, 222, 224; restart approval, 224, 225, 229; TEPCO relations, 51, 76; U.S. relations, 78, 84, 97, 98, 100

Nuclear Emergency Preparedness Act (Japan). *See* Act on Special Measures Concerning Nuclear Emergency Preparedness (Japan)

Nuclear Energy Institute (NEI), 94, 136, 196, 200, 205, 209; lobbying expenditures, 287n1; NRC relations, 116, 129, 232; predecessor, 189; SOARCA involvement, 212–13; writes own rules, 235, 236, 238, 239

nuclear fission. *See* fission

nuclear fuel, 5; cooling, 7, 8, 18, 158; damage, 56, 62, 71, 103, 126, 127–28, 145, 158; hydrogen generation, 7, 56, 145; reprocessing, 240; rods, 5, 7, 145;

supply, 171. *See also* fuel assemblies; dry cask storage; meltdown; spent fuel pools

Nuclear Information and Resource Service, 256

nuclear plant defense-in-depth approach. *See* "defense-in-depth"

nuclear plant inspections, 260, 266; IAEA, 50, 119; Japan, 16, 17, 48, 49, 51, 170, 223; NRC, 152, 173, 175, 199, 218, 235–36, 252, 253, 287nn4–5, 289n2; by plant owners, 191; precluded by high radiation, 180

nuclear plant licensing. *See* licensing of Japanese nuclear plants; licensing of U.S. nuclear plants

nuclear plant safety reports and rankings, 116, 152

nuclear plants, European, 224. *See also* Chernobyl nuclear accident, 1986

nuclear plant shutdown, 151, 201, 275; in emergency, 5, 7, 114–15; Hamaoka, 170; Japan-wide, 171–72, 222–24, 240; Three Mile Island, 142, 144. *See also* cold shutdown; restart of reactors

nuclear plant siting and construction, 40–41, 42, 43, 44, 153

nuclear plants, Japanese. *See* Fukushima Daini nuclear plant; Hamaoka nuclear plant; Kashiwazaki-Kariwa nuclear plant; licensing of Japanese nuclear plants; Monju nuclear plant; Ohi nuclear plant

nuclear plants, U.S., 40, 63, 94, 111–15, 166–70, 182–204, 212–13, 287n4; California, 34, 94, 98, 115, 128; Hurricane Andrew survival, 249; Minnesota, 247–48; near disasters, 174–76; New York, 40, 115–16, 198–99, 208, 209, 214, 287n6, 288n2; spent fuel storage, 83, 84. *See also* Browns Ferry Nuclear Power Plant; Davis-Besse Nuclear Power Station; Oconee Nuclear Station; Peach Bottom Nuclear Generating Station; Surry Nuclear Power Plant; Three Mile Island Nuclear Generating Station nuclear reactor design, 14, 41, 114, 176, 245–46. *See also* boiling water reactors (BWRs); Fukushima Daiichi design and construction; modular reactors; pressurized water reactors (PWRs)

Nuclear Regulation Authority (NRA), 240–41

Nuclear Regulatory Commission (NRC), 12, 15, *35,* 173, 182–203, 252–61, 287nn4–6; acquiescence to industry, 236–39; Advance Notice of Proposed Rulemaking, 188–90, 286–87n1; Advisory Committee on Reactor Safeguards, 153, 197–98, 214–15, 216, 217, 218; AEOD, 151; auditing, 49; backfit rules, 15, 191–98, 234, 259; citizens' suggestions to, 101–2; cover-ups and information withholding, 206–8, 213–14, 219–20; CRAC2 study, 207–9, 211, 212, 215, 216, 219, 220, 288nn8–9; disaster monitoring and assistance, 34–37, 54, 61–65, 77–82, 84–96, 98–101, 124–25, 129–30, 131; Fort Calhoun relations, 175; Individual Plant Examination (IPE) program, 191, 196, 199; infighting and dissent, 172–73, 176–77, 194, 196, 217, 218, 232–33; informed by *Wall Street Journal,* 136; Japan asks for assistance, 99, 131; lawsuit against, 194; McGaffigan role, 209–10, 211, 288n4; MELCOR, 139, 140, 264, 272, 286n3; NARAC relations, 138, 139; Near-Term Task Force (NTTF), 166–70, 172–73, 176–77, 185, 232–34, 237, 252–53, 254, 256, 260; NUREG-1738, 211; Office of Nuclear Reactor Regulation, 148, 218; Office of Nuclear Regulatory Research, 116, 288n5; Office of Nuclear Security and Incident Response, 129; Office of Research, 212, 217, 285n3; performance-based regulation, 235–39; Protective Measures Team, 62–63, 79, 125; public relations, 64–65, 101–2, 111–16, 148, 209; Regulatory Information Conference (RIC), 204–6, 219–20, 260–61; role in events of U.S. radiation release, 283–84n5; at Senate hearings, 182, *183,* 185; Severe Accident Policy Statement, 190, 195, 201; State-of-the-Art-Reactor Consequence Analyses (SOARCA), 130, 204, 205–6, 211–21, 268, 288n9, 289n1; Three Mile Island involvement and follow-up, 145, 147–48, 149, 151, 152, 153; "vulnerability assessment," 209. *See also* RASCAL (Radiological Assessment System for Consequence Analysis)

Nuclear Safety Commission (NSC), 15–16, 31, 38, 41, 55, 59, 108–9, 202–3, 225, 229, 241; corruption, 47; receives restart recommendations, 224; rules on stress tests, 229

Nuclear Sufferers Life Support Team, 157

Nuclear Utility Management & Resources Council, 196

NUREG-1738, 211

Obama, Barack, 92, 96, 128–29

Oconee Nuclear Station, 184–85, *184*

Ofunato, Japan, *53*

Ohi nuclear plant, 224, 229–31

oil reliance, 171, 228–29

oil spills, 102

Okuma, Japan, 16, 24, 40, 178, 243

Omaha Public Power District, *174,* 175

Osaka, 230, 231

Ostendorff, William C., 173, 182, *183,* 187, 233, 234, 278

overconfidence. *See* complacency and overconfidence

Oyster Creek Nuclear Generating Station, 40, 195

Pacific Ocean, 7, 19, 33, 34, 60, 128, 133, 138, 159; contamination, 155–56, *156.* *See also* seawater

Pacific Plate, 2–3

paperwork, 49, 67–68. *See also* auditing; falsification of reports

Pardee, Charles "Chip," 174, 239

passive cooling, 83

"passive" reactor systems, 245–46

Peach Bottom Nuclear Generating Station, 213, 214, 215, 216, 217, 218–19, 220, *237,* 288n6; Exelon plan for, 236–38

Pennsylvania government, 147, 148

Pennsylvania nuclear accident, 1979. *See* Three Mile Island nuclear accident, 1979

Pennsylvania nuclear plants. *See* Limerick Nuclear Power Plant; Peach Bottom Nuclear Generating System; Three Mile Island Nuclear Generating Station

Pietrangelo, Tony, 94–95, 278

plate tectonics, 2–3

plutonium, 171, 240

plutonium-239, 127, 128

portable generators, 25, 69, 169, 173, 214, 217

potassium iodide, 129, 138, 139, 256, 273

power outages, 8, 10, 12–13, 17, 18, 29, 31, 94, 124, 137, 251; computer simulations and hypothetical scenarios, 87–88, 218–19, 220, 246, 268; effect on news transmission, 117; effect on safety devices, 200; effect on SPEEDI, 58; effect on spent fuel pools, 81; planning for, 167, 168, 195–96, 203, 214, 217–18, 236, 237; U.S. plants, 175–76, 185, 249, 282n6

Prairie Island Nuclear Generating Plant, 247–48

pregnant women, 89, 148, 241

President's Commission on the Accident at Three Mile Island. *See* Kemeny Commission

press conferences, 26, 59–60, 92, 109; Three Mile Island, *149*

press releases: Met Ed, 147; NRC, 64, 80, 114; TEPCO, 18, 22, 23, 103, 107; U.S. Embassy, 88

pressurized water reactors (PWRs), 141, 142, 200, 213, 230, 273, 284n6

Price-Anderson Act, 246

probabilistic risk assessment (PRA), 192–93, 201, 202, 208–9, 213–14, 234, 273, 287n6, 289n1, 289n7; Indian Point, 288n2; mooted by Fukushima, 220–21

Protective Action Guides (PAGs), 80, 274, 283n3

protective clothing, 22, 26, 29, 31, 59, 75, 85, 96, 119, 181; worn by residents, 162

protests, 136, 161, 162, 165–66, 181, 222, 223–24, *223*, 231, 240; U.S. (c. 1970s), 146

psychological impact, 111, 207

Public Citizen, 152

public disclosure and participation, 257–58, 259, 260, 286–87n1, 287n3. *See also* secrecy

public fears, 100, 130, 148, 178, 186, 205, 209, 215, 222; information withholding in view of, 108, 118, 216

public health. *See* health consequences

public opinion, Japanese, 41–42, 46, 162, 222, 231, 239. *See also* protests

public relations, 63, 64–65, 101–2; Japanese government, 23, 26, 59, 103, 104, 108–9, 157; TEPCO, 107–8. *See also* Nuclear Regulatory Commission (NRC): public relations; press conferences; press releases

radiation detectors, 68, 144–45. *See also* dosimeters

radiation exposure, 31, 62–63, 78, 120, 129; CRAC2 on, 207; dose limits, 76, 78, 98, 158, 162, 177; physiological effect, 27–28; risk assessment, 192; Unit 3, 158; U.S. guidelines, 284n8. *See also* health consequences; lethality; protective clothing

radiation levels, 69, 85, 90, 91, 108, 122, 163, 177, 179; aerial, 69, 96, 99; California, 128; citizen monitoring, 109; government advice to public, 104–5, 162; "hot spots," 118, 119, 140, 159, 160, 178; Iitate, 118, 160; mapping, 98, 124; at plant gate, 26, 32, 82, 107; at sea, 68, 88, 155–56; south of Tokyo, 78; Three Mile Island, 144–45; Unit 2 and environs, 29, 90, 100, 158; Unit 3 and environs, 67, 72–73, 97, 99, 100, 158; Unit 4 and environs, 72, 78, 82; units of measure, 27; U.S. monitoring, 114. *See also* dosimeters; evacuation of residents; radiation exposure: dose limits

radioactive fallout. *See* fallout

radiation release, 7, 14, 28, 113, 118, 146, 155–56, 267–68; announced to public, 26; ASME view, 290–91n3; assumptions and forecasts, 41, 88, 140, 190, 205–6, 211, 213, 214, 215, 220; computer modeling, 58, 63, 79, 87, 288n9; design shortcomings, 17, 20,

58; hypothetical, 71, 78, 219; United States, 45, 146, 148, 150, 283–84n5. *See also* cleanup; public fears; radiation detectors; radiation exposure; source term; ventilation

radiation syndrome, acute. *See* acute radiation syndrome

radioactive gases, 5, 7, 32, 71, 74, 127–28, 146, 266–67; scrubbing, 20, 267; venting, *4*, 28, 169, 254

radioactive isotopes (radionuclides), 126–28, 155–56, 207, 211; released at Three Mile Island, 146. *See also* cesium-137; iodine-131

radioactive water, 155–56, 158–59, 164

RASCAL (Radiological Assessment System for Consequence Analysis), 63, 79–80, 87, 126, 129, 134, 207, 274

RCIC systems. *See* reactor core isolation cooling systems (RCIC)

reactor containment, 5–6, 14, 48, 70–71, 128, 209; breach of, 7, 26, 28, 30, 57, 71, 74, 77–79, 87–88, 98, 128, 158, 180, 186, 188, 195, 219; enhancement, 142, 188–89, 245; Three Mile Island, 145–46. *See also* Mark I containment; Mark II containment; Mark III containment

reactor core isolation cooling systems (RCIC), 8, 12, 218, 274; Unit 2, 17, 18, 24, 26, 62, 73, 265–66; Unit 3, 65, 66, 69, 266; United States, 217

reactor core meltdown. *See* meltdown

Reactor Safety Study (AEC), 192, 214

Reagan, Ronald, 191

"reasonable assurance of adequate protection." *See* "adequate protection"

redundant systems. *See* backup systems

Regulatory Information Conference (RIC), 204–6, 217, 219–20, 260–61

relocation of residents. *See* evacuation of residents

rem (unit of radiation), 27, 274

residents' evacuation. *See* evacuation of residents

restart of reactors, 46, 151, 224, 225, 227, 228, 229; Hamaoka, 170–71; Kashiwazaki-Kariwa, 228, 239; Monju, 286n1; Noda promises, 172, 223; NRA

role, 241; Ohi, 224, 229, 230–31; Three Mile Island, 142

retrofitting. *See* backfitting

RIC. *See* Regulatory Information Conference (RIC)

Richter, C. F., 43

Richter scale, 281n1

risk analysis, 191–97, 201, 234, 253, 259. *See also* probabilistic risk assessment (PRA)

Rokkasho Reprocessing Plant, 240

Roberts, Thomas, 197

Ronald Reagan (ship). *See* U.S.S. *Ronald Reagan*

Roos, John V., 69, 85, 87, 88–89, 125, 132, 134, 278

Safety Commission. *See* Nuclear Safety Commission (NSC)

safety reports and rankings. *See* nuclear plant safety reports and rankings

Sakurai, Katsunobu, 106–7

salt, 100, 131, 134, 135, 137, 286n1

saltwater. *See* seawater

SAMA. *See* severe accident mitigation alternatives (SAMA)

SAMGs. *See* severe accident management guidelines (SAMGs)

sand, 88, 90, 91, 98

sandbagging, 37, 175

Sandia National Laboratories, 134, 264, 265, 286n3; CRAC2 study, 207–9, 211, 212, 215, 216, 219, 220; SOARCA analysis, 217

San Onofre Nuclear Generating Station, 115, 128

Sanriku Coast, 9–10, 40, 51, 52, 53

Santiago, Patricia, 220

scale. *See* earthquakes: magnitude scales; instrument scale problems; International Nuclear and Radiological Event Scale (INES)

scandals, 46, 48, 50

scenarios. *See* "best-estimate" scenarios; worst-case scenarios

Schaperow, Jason, 220

schools and schoolchildren, 148, 160, *161*, 162

Scidmore, Eliza Ruhamah, 282n4

Science and Technology Agency. *See* Japan Science and Technology Agency

Scranton, William, III, 148

seawalls, 10, 53, 231, 246, 250

seawater: emergency use, 10, 33, 57, 60–62, 65, 67, 72, 77, 79, 86, 97–100, 107, 121, 123–24, 131, 137, 206, 265; normal use in cooling, 7, 19, 40–41, 52

secrecy, 40, 51, 151, 152, 179, 184, 212, 215, 219–20, 257; of inspections, 223. *See also* classified information

Seismic Isolation Building, 4, 9, 37, 75, 80–81, 85–86

seismology, 2–5, 41, 42–44, 51, 63, 114, 229. *See also* fault lines

Self-Defense Forces (SDF), 21, 22, 96, 97, 98–99, 118, 121, *122*, 285n5

Senate Environment and Public Works Committee, 94, 182

Sendai, 1, 3, 5

September 11, 2001, terrorist attacks (9/11), 95, 173, 208–10, 211, 235, 241

severe accident management guidelines (SAMGs), 200, 202, 215, 275, 287n4

severe accident mitigation alternatives (SAMA), 259–60

shareholders' meetings, 165–66

Sheron, Brian, 64, 134, 217, 278

Shimizu, Masataka, 21, 25, 50–51, 67, 73, 76–77, 107, 277; resignation, 165, 166

Shimokobe, Kazuhiko, 227–28, 277

sievert (unit of radiation), 27, 275

simulated accidents. *See* computer simulation and modeling; emergency drills

Smith, Aileen Mioko, 110, 277

SOARCA. *See* "State of the Art Reactor Consequence Analyses" (SOARCA)

soil contamination, 120, 160, 162–63, 178, 214, 254, 256

source term, 126–30, 132, 133, 137–40, 189–90, 206, 214, 216, 288n8; definition, 275

South Korea, 68, 160

Soviet Union, 39. *See also* Chernobyl nuclear accident, 1986

SPEEDI. *See* System for Prediction of Environmental Emergency Dose Information (SPEEDI)

Speis, Themis, 197

spent fuel, dry cask storage of. *See* dry cask storage

spent fuel pools, 7, 70–71, 76, 81, 83–84, 88, 100, 121, 123–24, 131, 132, 139, 158, 159, 163, 174, 256, 275, 289–90n8; instrumentation, 236; in PWRs, 284n6; risks, 210–11; Unit 1, 57, 70, 88, 89, 90, 93, 121–22, 137; Unit 2, 70, 88, 89, 90, 93, 121–22, 137, 267; Unit 3, 70, 78, 88, 89, 90, 91, 93, 99, *122*, 137; Unit 4, 69–71, 81, 82, 87–93, 95–96, 121, *123*, 133, 136–37, 255; United States, 255

State Department. *See* U.S. State Department

state governments, 64, 147, 148, 283–84n5

State-of-the-Art Reactor Consequence Analyses (SOARCA), 130, 204, 205–6, 211–21, 268, 275, 285n3, 288nn9–10, 289n1

station blackout. *See* power outages

steel: in reactor containment, 5–6, 14, 48, 113, 188; in spent fuel storage, 7, 70, 83; vulnerability in cases of meltdown, 7, 113, 195, 219

storage of contaminated materials. *See* disposal and storage of contaminated materials

Strauss, Lewis L., 39

"stress tests," 170–71, 224–25, 229, 230–31, 259

strontium-90, 127

Stutzke, Martin, 287n6

Sugaoka, Kei, 48

Sullivan, Randy, 216

Sumatra earthquake, 2004, 47

Surry Power Station, 213, 214, 215, 216, 218

Svinicki, Kristine L., 167–68, 172–73, 182, *183*, 234, 288

Sweden, 189, 254

System for Prediction of Environmental Emergency Dose Information (SPEEDI), 58–59, 63, 118, 207, 275

Takekuro, Ichiro, 60, 61, 277

taxation, 40, 178, 204, 226, 230

television, 1, 10, 12, 33, 54, 55, 59, 62, 85,

91, 97, 171, 180; Edano on, 103; United States, 115–17

tellurium, 109, 118, 156

TEPCO. *See* Tokyo Electric Power Company (TEPCO)

terrorist attacks (hypothetical), 183–84, 209, 211, 214, 241

terrorist attacks, September 11, 2011. *See* September 11, 2001, terrorist attacks (9/11)

Thornburgh, Richard, 145, 148

Three Mile Island nuclear accident, 1979, 14–15, 44–45, 101–2, 140–54, 188, 202; cleanup, 146, 164; industry effect and response, 58, 61, 112, 168, 205; INES rating, 103–4; NRC effect and response, 152, 187, 188, 195, 196, 200, 232, 253. *See also* Kemeny Commission

Three Mile Island Nuclear Generating Station, *143*

thyroid gland, 27, 99, 127, 129, 133, 134, 138, 139; potassium iodide and, 273

tidal waves. *See* tsunamis

Tohoku Earthquake, 2011, 1, 3–5, 9, 10, 13, 47, 282n5

Tokai, 43

Tokaimura research center, 40; nuclear accident, 1999, 37–38, 103

Tokyo: earthquake vulnerability, 229; protests, 223, *223*; radiation threat to, 77, 78, 79, 89, 90, 124, 133, 134, 138, 139, 140, 256

Tokyo Electric Power Company (TEPCO), 48–54, 61; blamed, 244; blame-placing by, 103, 180, 244; communications, 76–77, 107–8, 264; compensation plan, 165, 289n4; cover-ups and falsification, 48–49; denial of "owning" radiation, 179; disaster response, 12–13, 16–26, 30, 55, 59, 60–61, 85, 86, 88, 100; finances, 165, 178–79, 226–28; Hatamura committee criticism, 226; incestuous practices, 47, 48; media coverage, 107; new management, 227–28, 239; NRC relations, 125, 135, 136, 155, 163; openness to help, 100, 102, 131, 134–35; postmortem arguments, 246–47, 266–67; power outage preparation, 203; pre-disaster retrofitting, 202;

press conferences, 67; protests against, 161; public relations, 107–8, 264; radiation readings, 97; recovery plan, 163–66; repowering efforts, 97, 99; self-assessment, 180; shareholders' meetings, 165–66; siting and construction, 40, 41, 42; tsunami planning, 51–54; water contamination problem, 158, 159–60. *See also* Fukushima Daichi nuclear plant

tornadoes, 188, 192, 249, 253

torus. *See* wetwell (torus)

Toshiba, 164, 290n10

Trapp, Jim, 77, 78–79, 80, 84, 85, 86, 136, 278

Trautman, Stephen, 88, 278

travelers, American, 93, 99–100

tsunami (word), 282n4

tsunamis, 9–13, *11*, 22–23, 229, 247–48; damage from, *53*, 63, 159, 250; effect on fishing industry, 156; evacuation after, 57–58; fatalities, 98; Indian Ocean, 47; Minamisoma, Japan, 106–7; planning for, 51–54, 171, 246; search for victims, 181; TEPCO view, 180; U.S. warnings, 34

turbine buildings, 10, 30, 53, 56, 62, 158, 160, 276

turbine generators, 5, 8, 48, 56

Turkey Point Nuclear Generating Station, 249

Ulses, Anthony "Tony," 69, 77, 78–79, 80, 84, 85, 86, 95, 278

uniforms, 284–85n2. *See also* company uniforms

Union of Concerned Scientists, 94, 106, 193, 201, 206, 283n1, 286n1; lawsuit by, 194

United States: aid from sought, 77; Atoms for Peace, 39–40; earthquake hazards, 113, 114; generation equipment in Japan, 282n8; western states, 64, 99, 113, 114, 128, 138, 139. *See also* Alaska; California; nuclear plants, U.S.; Nuclear Regulatory Commission (NRC); Three Mile Island nuclear accident, 1979; White House

uranium, 7, 37, 171; in fuel pellets, 5, 271

U.S. Air Force, 97

U.S. Atomic Energy Commission. *See* Atomic Energy Commission (AEC)

U.S. citizens. *See* Americans

U.S. Congress, 36, 92, 116, 187, 205, 246, 247; Atomic Energy Act, 187; EPAct, 204–5; hearings, 85, *92*, 93–94, 182, *183*, 185; Three Mile Island investigation, 149, 193

U.S. Defense Threat Reduction Agency. *See* Defense Threat Reduction Agency

U.S. Department of Defense, 134, 138

U.S. Department of Energy (DOE), 63–64, 68, 133, 134, 137, 139; collaboration, 102, 131, 288n5; Office of Naval Reactors, 68, 88, 89; Radiological Assistance Program, 84, 97; simulations, 99. *See also* National Nuclear Security Administration; Sandia National Laboratories

U.S. Department of Homeland Security, 99–100, 114

U.S. Department of State, 80, 92–93, 139

U.S. Embassy, Tokyo, 61, 62, 69, 80, 100, 129, 132; contingency planning, 124; fears within, 87, 89, 133; NRC at, 77, 84, 86, 100, 125

U.S. Environmental Protection Agency (EPA), 80, 114, 126, 129, 132, 138, 139, 284n8

U.S. Fleet Activities Yokosuka, 78, 79–80

U.S. Geological Survey, 4, 44, 114, 282n5

U.S. military, 68, 69, 77, 78, 88, 97, 121, 138, 160

U.S. Navy, 68, 88, 160

U.S. Nuclear Regulatory Commission. *See* Nuclear Regulatory Commission (NRC)

U.S.S.R. *See* Soviet Union

U.S.S. *Ronald Reagan,* 68

vague language: in guidelines and standards, 48, 187, 190, 191, 194, 224, 290n2; from sources on Fukushima events, 22, 98, 130

van Loon, Eric, 206, 208

ventilation, 19–20, 236, 238, 251, 276; filters, 128, 188, 231, 254, 255, 268; government approval, 21–22, 30, 31; NRA measures, 241; planning for, 169, 246; policy recommendations, 197–98; Three Mile Island, 146, 148; Unit 1, 19–20, 22, 24–26, 28–32, 57; Unit 2, 25, 73–74; Unit 3, 65, 66–67, 69. *See also* exhaust stacks; hardened vents

video, 12, 16, 54, 61, 95; conferencing, 25, 86

Virgilio, Martin, *86,* 278; event monitoring (days 2–4), 61–62, 63–64, 77; event monitoring (days 5–8), 81, 85, 87, 89, 98–99, 102; event monitoring (days 9–10), 125, 132–33, 134, 138; response to NTTF report, 233

Virginia nuclear plants. *See* North Anna Power Station; Surry Power Station

Vogtle Electric Generating Plant, 245, 282n6, 288n1

voluntary industry measures, 154, 170, 173, 198, 199–200, 202–3, 235–36, 253; Japan, 51–52, 157, 246–47. *See also* FLEX program; severe accident management guidelines (SAMGs)

von Hippel, Frank, 192

Wagner, Brian, 36

water, fresh. *See* freshwater

water, ocean. *See* seawater

water cannons, 90, 96, 97, 121, *122*

water contamination. *See* radioactive water

Waxman, Henry, 93

weather, 128, 207–8, 214, 219, 221, 249. *See also* hurricanes; tornados; wind

Weber, Mike, 138, 278

Weiss, Ellyn, 194

western states (U.S.). *See* United States: western states

Westinghouse Electric, 39, 200, 290n10; AP1000 reactor, 114, 244, 245

wetwell (torus), 6, 19, 62, 73, 75, 136, 267–68, 276, 284n7; breaches, 77–78, 79, 266; scrubbing in 20, 28, 267; ventilation, 238

White House, 92, 111, 129–30, 137–39, 140, 205; NRC relations, 61, 89, 129, 130, 132, 133, 137–39, 166, 177; Office of Science and Technology Policy, 133, 267; Three Mile Island involvement, 147, 148. *See also* Kemeny Commission

Wiggins, Jim, 124, 133, 278
Wilson, George, 168
wind, 55, 58, 77–80, 82, 87, 97, 118, *119*,
126, 128, 152, 156, 207; downwind
fatalities, 211; Hurricane Andrew, 249;
Tokyo and, 77, 78, 79, 89, 90, 133, 134
World Nuclear Association, 35
worst-case scenarios, 125, 126, 129–30,
132, 133, 138, 187; in CRAC2, 208,
215, 288n8; defense-in-depth approach
to, 248; discussed in closed NRC
briefing, 216; ignored in FLEX plan,
238

xenon-133, 146

Yokosuka naval base. *See* U.S. Fleet
Activities Yokosuka
Yokota Air Base, 97, 98
Yomiuri Shimbun, 46, 47
Yoshida, Masao, *21,* 243, 277; death, 284n1;
insubordination by, 60–61, 86; Kan
relations, 25, 31; March 11 management,
10, 20–21, 23–24, 25–26, 30–31, 32, 33,
37; March 12–14 management, 37, 55,
57, 60–61, 65–68, 73, 74, 75, 76, 80–81,
85–86; post-disaster statements, 13, 85
Yoshizawa, Atsufumi, 243, 277

zirconium, 5, 6, 7, 88, 145; ignition of, 71,
82, 99, 211

PUBLISHING IN THE PUBLIC INTEREST

Thank you for reading this book published by The New Press. The New Press is a nonprofit, public interest publisher. New Press books and authors play a crucial role in sparking conversations about the key political and social issues of our day.

We hope you enjoyed this book and that you will stay in touch with The New Press. Here are a few ways to stay up to date with our books, events, and the issues we cover:

- Sign up at www.thenewpress.com/subscribe to receive updates on New Press authors and issues and to be notified about local events
- Like us on Facebook: www.facebook.com/newpressbooks
- Follow us on Twitter: www.twitter.com/thenewpress

Please consider buying New Press books for yourself; for friends and family; or to donate to schools, libraries, community centers, prison libraries, and other organizations involved with the issues our authors write about.

The New Press is a 501(c)(3) nonprofit organization. You can also support our work with a tax-deductible gift by visiting www.thenewpress.com/donate.